‘Every bright young mind in Australia should read *Adam Spencer's Big Book of Numbers* – and we oldies would benefit, too.’

Peter FitzSimons

‘Excellent, can't wait to geek-out my dinner party guests with my new maths facts.’

Maryanne Demasi, *Catalyst*, ABC Television

‘Adam Spencer's love of numbers is legendary – it leaps off the page and will have readers spellbound all the way from 1 to 100.’

John Watkins, former NSW Education Minister

‘Adam Spencer's TED talk at Longbeach California has had over a million views and is the stuff of legend. This book shows why. One of the great maths communicators.’

Janne Ryan, Founding Executive Producer, TEDxSydney

‘Numbers are fascinating and you don't have to like maths to be spellbound by them. Adam Spencer knows numbers and has invested them with love.’

Rodney Cavalier, former NSW Education Minister

‘Adam Spencer engages children with the history, the mystery and the wizardry of mathematics. Any child of any age reading this book will have their interest in numbers multiplied and mathematics will have their undivided attention.’

Eve Field, Principal, Kurri Kurri Public School

‘This delightful, tantalising and timely book ... gives you both the fun and surprise as well as the *real stuff* on how to use maths as the basis of everything.’

Robyn Williams

Adam Spencer's

BIG BOOK *of* Numbers

EVERYTHING YOU WANTED TO KNOW ABOUT THE NUMBERS 1 TO 100

XOUM PUBLISHING

Sydney

First published in Xoum by Brio Books in 2014
This edition published in 2016
Reprinted in 2018, 2019

Brio Books Pty Ltd
PO Box Q324, QVB Post Office,
NSW 1230, Australia
briobooks.com.au

ISBN 978-1-925143-13-3 (print)
ISBN 978-1-921134-33-3 (digital)

Cataloguing-in-Publication data is available from the National Library of Australia

Endpaper design: Penrose Tiling of the Plane (see page 142)
Cover design, text design and typesetting: Xou Creative, xou.com.au
Illustrations: Roy Chen, Xou Creative
Author photograph: David Stefanoff
Printed and bound in China through Asia Pacific Offset

This book is dedicated to a number and ... a mysterious woman ...

To the number 4 where this whole crazy adventure began.

The first number I obsessed over, wanted to get inside, pick apart and understand. I can remember putting things in groups of 4, which was two groups of 2, doubling 4, and doubling you again and again up to numbers too fantastic to think I'd ever really understand – one hundred ... maybe even a thousand. Not big numbers now, but I was only little at the time. I think I was about ... four.

I'll never know why it was you and not 3, a thoroughly dignified prime number, or 6, a perfect number, or 7 ... actually, I know why it wasn't 7 – overrated, for mine. But that's another story.

Four, you're a keeper.

This book is also dedicated to a mystery woman who would now be nearing 30. I had just started doing breakfast radio on triple j when you, as a first-year high school student (and exactly the age of what I hope will be many of the readers of this book), sent me a message on the breakthrough technology of email which simply said:

'Since you've been on triple j, talking about your sort of stuff ... I just get picked on a lot less at school.'

This book is for you and for everyone who, like me, loved 4 and 3 ... what the hey, even 7. They're a pretty lovely bunch and it's a thrill to bring them to you.

INTRODUCTION

I love numbers. I have for as long as I can remember. I explain it to people by saying that numbers are the musical notes upon which the symphony of the universe is played, or that they are the jigsaw pieces of the puzzle of life, but more than anything, I think they are simply beautiful.

Now, if you're holding this book, you fall into one of two groups (see, even when I write an introduction to a book, I write like a mathematician). Either you love numbers as much as I do, or you need some convincing.

If you love numbers, dive straight on in. Immerse yourself in the beautiful facts, patterns and problems. Try to solve the mysteries and use what is here as a springboard into the oceans of amazing mathematics that are out there.

If you need a bit of convincing about the beauty of numbers, this is the book for you. Tread more sedately and stop to sniff the digits. I truly believe you'll find joy in this book.

A lot of this is stuff adults never saw in high school, nor will current school students ever see, but it's set out in a way that even though some of it is second- or third-year university material, it's not that hard to get your head around.

Please, if you see something and it seems a bit too complicated, feel free to skip it and come back and have a look again later. Just roll onto the next chapter and enjoy the maths and other weird and wonderful facts therein.

When you're enjoying the *Big Book of Numbers* if there's anything you'd like to ask, go to adamspencer.com.au and let me know. Ask questions, start arguments, even slightly ruin my day by pointing out any mistakes you've found. I'd love to hear from you.

Strictly speaking, this book is an 'all improved second edition' which builds upon a book that I wrote in, if I remember correctly, 1998. Honestly so long ago I can't remember exactly when.

Since then, I've learnt a lot more, got a bit more ambitious, the internet has really come along, and I've read some incredible books. In particular, I've been inspired by Martin Gardner's beautifully-titled *Are Universes Thicker Than Blackberries?*, David Well's *Penguin Dictionary of Curious and Interesting Numbers*, Rob Eastaway's *How Many Socks Make a Pair?*, Derrick Niederman's *Number Freak*, Ivan Moscovich's puzzle masterpieces, Clifford Pickhover's *The Math Book*, David Darling's *Universal Book of Mathematics* and the hilarious and informative British television juggernaut *Qi's* fact books. Last but not least, mathworld.wolfram.com is now up and running and is the most exquisite resource to find out anything else you want to know about numbers.

If my *Big Book of Numbers* can inspire and entertain you as much as these cracking reads have me, I'll be one happy little mathematician.

Thanks for buying the book – enjoy.

Adam S

NUMERO UNO

It's not strictly correct to say that 1 is the 'first number', because there are negative numbers, zero, and an endless list of other numbers that are 'less than' 1. But 1 certainly is the first number most of us encounter and it's probably the first number of which anyone ever thought. That realisation – that we can have 1 of something, or 2 of them, or small bits of things that aren't really 1 thing – is where this whole wacky caper called mathematics started.

So, you multiply any number by 1, you get the number that you started with. In fancy-pants mathematics talk we say $a \times 1 = a$ for any number a. For this reason 1 is called the 'multiplicative identity'. Similarly, $a + 0 = a$ for all a and so 0 is called the 'additive identity'.

BENFORD'S LAW

Electrical engineer and physicist Frank Benford noticed something pretty amazing about naturally occurring sets of numbers. If you take 500 cities from around the world and list their populations, or all of the countries in the world and list their areas, or look at the front page of your local newspaper every day for 5 years and note every number that occurs on those pages, these lists follow an interesting pattern.

You might think that the first digit in any of these numbers is just as likely to be a 3 as it is a 7 or a 9 or whatever. But no! Benford's Law says that the first digit will be a 1 about 30.1% of the time, a 2 about 17.6% of the time, a 3 (12.5%), a 4 (9.7%) and so on, down to poor old 9 which will only start about 4.6% of numbers on the list.

Among many cool applications, Benford's Law is used to catch out people who submit dodgy tax returns. If the list of all the numbers in your tax return doesn't follow a Benford Distribution, your tax return may well arouse suspicions!

Take a 1-metre piece of string and cut it in half.

Cut one of those halves in half again and cut one of those two pieces in half ... and so on ...

Our 1-metre piece of string starts to look like this:

... and so on. So the 1-metre piece of string is now made up of pieces:

$\frac{1}{2}$ m, $\frac{1}{4}$ m, $\frac{1}{8}$ m, $\frac{1}{16}$ m ... long, which illustrates what we call an infinite series, the equation:
$1 = \frac{1}{2} + \frac{1}{4} + \frac{1}{8} + \frac{1}{16} + \frac{1}{32} + \ldots$ an infinite number of terms, in this case, with a finite sum!

THE BASICS

If you stick with this book for any period of time, there are going to be a few things that at first might confuse you ... or even freak you out. So let's start the freaking out process here.

One is *not* a prime number.

When you read this you probably have one of two reactions. You're either thinking, 'Wow, why is that?' or you're sitting with your eyebrows scrunched up thinking, 'What's a prime number again?'

Let's start with the eyebrow scrunchers first.

Six is not a prime number because we can write it as the product of 2 smaller whole numbers, that is, $6 = 2 \times 3$. We call 2 and 3 'factors' of 6 and describe 6 as a 'composite' number.

For the number 7 we can write $7 = 1 \times 7$ but we can't find any other factors of 7 apart from 1 and itself. So 7 is a 'prime' number.

All good? Excellent. But once you understand what a prime number is, it's understandable to think of 1 as prime, after all, $1 = 1 \times 1$ and we can't break it into any other factors.

But try this on for size.

If we multiply any two prime numbers together, the answer we get obviously can't be prime. Five and 7 are prime, $5 \times 7 = 35$, and 35 is clearly composite.

Let's pretend that 1 is prime. We all agree 3 is prime. So multiplying these two primes together we get $1 \times 3 = 3$. We get 3 by multiplying two prime numbers together so 3 is composite. But we just said 3 is prime! It can't be both ... aaaargh, this is hurting my head.

Assuming 1 was prime has led to a situation that makes no sense. So 1 isn't prime. In fact, it isn't composite either. It's a special case due to 1 being the 'multiplicative identity'. Mathematicians call 1 a 'unit'. In the unlikely event someone asks you, zero is also neither prime, nor composite.

Using this sort of argument when you assume something is true and show it can't be is called a 'proof by contradiction'. It's one of the most powerful forms of thinking in all of mathematics.

2

Two can only be written as $2 = 1 \times 2$ so 2 is a prime number. Two is the only even prime number, because every other even number is divisible by 2, and so can't be prime.

IRRATIONAL NUMBERS

It's obvious that $3 \times 3 = 9$. We describe this as '3 squared equals 9' and write $3^2 = 9$.

The reverse of this process is to say that 'the square root ($\surd$) of 9 is 3', written $\sqrt{9} = 3$.

Sometimes square roots are whole numbers. For example, $\sqrt{25} = 5$, $\sqrt{49} = 7$. But most square roots aren't quite so 'clean'.

The square root of 2 is called an 'irrational' number. This doesn't mean that it can't be trusted to behave at parties, but that it can't be written as a fraction $\frac{a}{b}$ where a and b are whole numbers.

Way back in 500 BC, the philosopher Hippasus of Metapontum was the first person to show that $\sqrt{2}$ was irrational.

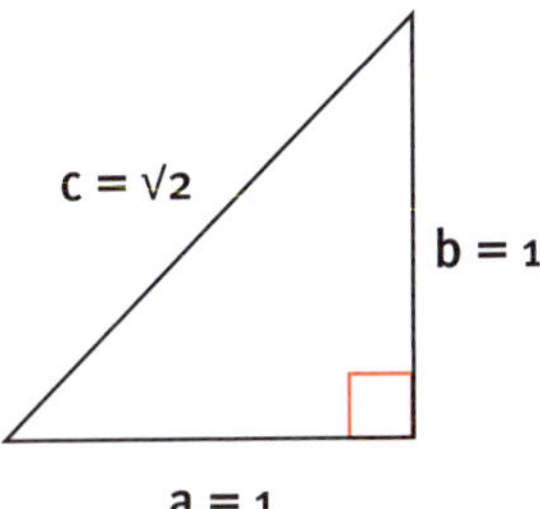

Gun mathematician and philosopher, Pythagoras, who also lived around 500 BC, discovered that in a triangle with a 90-degree angle (a right triangle) with shorter sides length a and b and longer side length c, the lengths are related to each other by the formula $a^2 + b^2 = c^2$. So in a right triangle with both shorter sides length 1, we calculate the hypotenuse or longest side length $1^2 + 1^2 = c^2$, giving $2 = c^2$ or $c = \sqrt{2}$.

Photograph of a Basile Bouchon automated loom, displayed at the Musée des arts et métiers, Paris. Source: Dogcow

Two is the basis of the 'binary system', a form of mathematics which uses only zeros and ones. It's the basis of the technology that drives computers, which we'll meet in chapter 64. In 1725 Frenchman Basile Bouchon invented a device consisting of a roll of perforated paper punched with holes that could control the threads on a mechanical loom. The same idea was used in the pianola. Then Charles Babbage used the system in his Analytical Engine (see chapter 79). Thanks to binary mathematics, you can do all sorts of amazing things with today's 'analytical engines' ... even take selfies.

1 + 1 = 2

Before you say, 'Yeah, thanks for that, Adam,' this is in fact a deep statement that, if you're being strict, takes tremendous effort to prove.

EVEN NUMBERS

Even numbers are numbers divisible by 2. Two is the first even number we usually encounter but, strictly, zero is even, as are –2, –4, –6 …

RED PLANET

Mars has 2 moons – Phobos and Deimos, children of Ares the God of war in Greek Mythology (the Romans knew Ares as Mars).

Curiosity Self-Portrait at 'Windjana' Drilling Site. Source: http://www.nasa.gov/jpl/msl/pia18390/#.U7oLv2eKJM9

EULER'S FORMULA FOR POLYHEDRA

Legendary Swiss mathematician and physicist Leonhard Euler made thousands of useful mathematical observations. He really was a complete gun. One of them was about 3-dimensional (3D) shapes. Euler realised that in any 3D shape with no notches or holes through it (which mathematicians call a convex 3-dimensional polyhedron), the number of vertices (corners), faces and edges obey the equation $V + F - E = 2$.

QUIZ QUESTION: Verify Euler's formula for each of these 3-dimensional shapes: a cube, a square pyramid and an octahedron.

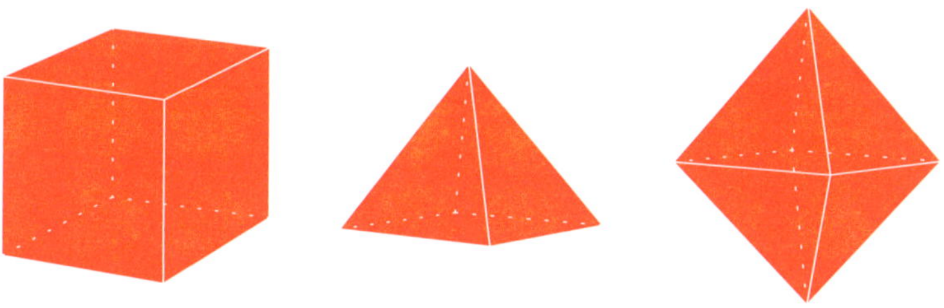

ANSWER AT THE BACK OF THE BOOK

PYTHAGOREAN TRIADS

It's clear that $3^2 + 4^2 = 9 + 16 = 25$, but $5^2 = 25$ too. So $3^2 + 4^2 = 5^2$. Looking back at Pythagoras' Theorem (in chapter 2), we see that there is a right-angled triangle with shorter sides 3 and 4 and longer side 5. We say (3, 4, 5) is a Pythagorean triad (or triple).

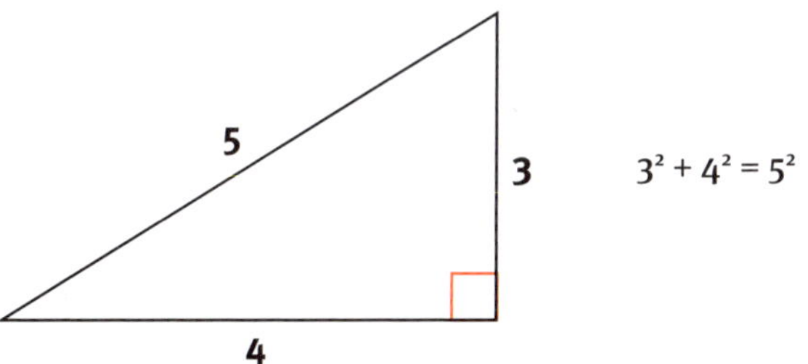

JOHNSON'S THEOREM

A neat little party trick, first noticed by the American mathematician Roger Johnson in 1913, is this – if you have 3 circles, all of the same size, that intersect at a common point:

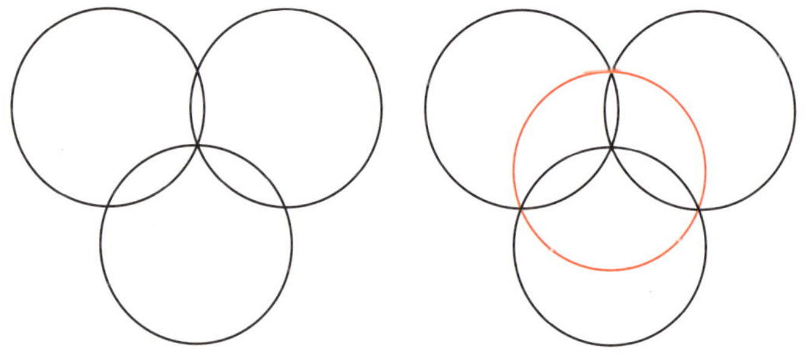

Then the other 3 points of intersection of the circles must lie on another circle of the same size.

This is known as Johnson's Theorem, but to be honest, while I did call it a 'party trick', that does depend on what type of parties you go to!

MORLEY'S THEOREM

To split an angle into 3 equal parts is to 'trisect' it. The brilliant American mathematician Thomas Morley realised that if you trisect the angles at each corner of a triangle the trisectors intersect, forming an equilateral triangle.

Regardless of the shape of the big triangle, this middle triangle will have 3 equal sides (equilateral).

The human middle ear consists of 3 bones – the malleus, the incus and the stapes ...

Three is a Fibonacci number. The Fibonacci numbers are: 1, 1, 2, 3, 5, 8, 13, 21 ... Can you see the pattern? You add 2 numbers in a row to get the next number. Fibonacci numbers are among the most famous and often-used numbers in mathematics.

... these are the smallest bones in your body ...

There's an easy way to check if a number is divisible by 3, without a calculator. If the digits of a number add up to a number divisible by 3, then the original number can be divided by 3. There are no exceptions. So, we know that 39,123 is divisible by 3 because $3 + 9 + 1 + 2 + 3 = 18$ and $18 = 3 \times 6$. In case you're wondering, $39{,}123 = 13{,}041 \times 3$. For much bigger numbers, you can repeat the process of adding the digits until you get a small enough sum that you can check for divisibility by 3.

... the stapes being the littlest.

There are 4 letters in the word 'four'. This is the only number for which the number of letters and the number itself match up.

MY FIRST LOVE

The first number I ever fell in love with was 4. I don't know exactly why, but the fact that you could double 1 a couple of times and get 4, that there were 4 dots in a square, and that you clapped in 4/4 time made 4 seem very natural and beautiful to me.

Even today when setting things like the volume of an MP3 player or the radio in the car I'll always settle on a multiple of 4.

Maybe it's just me?

$4 = 2 \times 2$

This doesn't seem like the greatest of revelations, but this diagram tells us a couple of very cool things about the number 4.

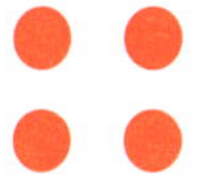

Four is the first number that can be written as the product of factors other than 1 and itself. So 4 is the first 'composite' number.

And because $4 = 2 \times 2$ we could write it as 4 dots in a square.

As we've seen, we say that 4 is '2 squared' and we write this $4 = 2^2$. The number floating around in the air is called a 'power' or 'exponent'. We also say 4 is '2 to the power of 2'.

Extending this notation, $5 \times 5 \times 5$ could also be written as 5^3 and called '5 to the power of 3'.

PARTITIONS OF 4

We can write the number 4 as 4 = 4 or 4 = 3 + 1 or 4 = 2 + 2 or 4 = 2 + 1 + 1 and so on. These are all called 'partitions' of 4. When we write partitions we consider 2 + 1 + 1 to be the same partition as 1 + 2 + 1. When writing out the partitions of a number we tend to write the numbers within the partition all smallest to largest or largest to smallest to make it easier to keep track of them and make sure you get every one.

QUIZ QUESTION: Find all 5 partitions of 4.

ANSWER AT THE BACK OF THE BOOK

THE FOUR 4S PROBLEM

Have a look at the following 4 equations:

$$1 = \frac{4 \times 4}{4 \times 4} \qquad 2 = \frac{4}{4} + \frac{4}{4}$$

$$3 = \frac{4 + 4 + 4}{4} \qquad 4 = (4 - 4) \times 4 + 4$$

One famous, fun and very frustrating maths problems is this: with only basic mathematics write all of the numbers from 1 to 100 using exactly four 4s.

When we say 'basic mathematics', you only need +, −, x, ÷ along with ! (see chapters 6 and 24), √ (see chapter 2) to get most of the numbers, and a couple more obscure signs to work out the toughest ones.

Using the basics including $4! = 24$, $\sqrt{4} = 2$ and using two 4s to get 44, you can make all the numbers up to 30 with four 4s, without too much trouble. Give it a go!

Deoxyribonucleic acid (DNA) is the amazing molecule that is crucial to the functioning of us and indeed all living things.

Our DNA is housed in the nucleus of our cells and packaged as long structures called chromosomes. For all its incredible complexity, at its heart are four – *just* four – basic building blocks called nucleotides.

The nucleotides are made up of a sugar called deoxyribose, a phosphate molecule and one of four bases, Thymine, Cytosine, Adenine and Guanine and it's the arrangements of billions of these A, C, G and Ts that packs the phenomenal wealth of information into our DNA.

Each cell contains 3 billion pairs of bases, which if uncoiled would give a strand of about 2 metres of DNA. The total DNA in your trillions of cells would stretch beyond the edge of our solar system.

We have come so far in the 50 years since Francis Crick, James Watson and the much lesser-known, but absolutely awesome Rosalind Franklin unveiled the double-helix structure of DNA. Yet we have so far to go. Who would have thought four little letters – A, C, G and T – could tell us so much?

5

Five members seems to be the preferred number for pop harmony boy bands. While they span different decades and hail from different countries, Backstreet Boys, 'N Sync, 5ive, Take That, New Kids on the Block, One Direction et al have a couple of things in common – their music is truly deplorable *and* they all have 5 members. As far as I can tell, the allocation of talent across the 5 members is traditionally 2 fantastic dancers; 1 guy who looks great in slacks; 1 moody, brooding loner (complete with nose ring and facial hair); and 1 guy who can actually ... sort of ... sing.

FIVE DIMENSIONS

You probably know that the area of a circle is 'pi r squared' which we write as πr^2. Can I push you a bit further and remind you that the volume of a sphere is $\frac{4}{3}\pi r^3$?

So when our radius is 1, the 2-dimensional 'unit circle' as we call it has area $A = \pi$. Similarly, the 'unit sphere' is a 3-dimensional object, which has volume $V = \frac{4}{3}\pi$.

Asking you to picture spheres in 4 dimensions or higher might well make your eyes bleed, but here is an amazing result. The unit spheres don't just keep getting bigger as the dimensions go higher. In fact, the volumes of the unit spheres from 2 dimensions up are:

$$\pi, \frac{4\pi}{3}, \frac{\pi^2}{2}, \frac{8\pi^2}{15}, \frac{\pi^3}{6}, \frac{16\pi^3}{105}, \frac{\pi^4}{24}, \text{etc}$$

You can see that while the powers of π get higher, the fraction gets smaller much more quickly. Surprisingly, the volume heads to zero as the dimensions head towards infinity.

So the 'biggest' unit sphere is the 5-dimensional one:

$$V = \frac{8\pi^2}{15}$$

Next time you're playing basketball in 5 dimensions point this out and really impress people.

Every number that ends in a 5 is divisible by 5 (all other multiples of 5 end in zero). Oh, and $5 = 1^2 + 2^2$ is the smallest number that can be written as the sum of two different positive squares.

PLATONIC SOLIDS

I'm sure you're familiar with a cube. To describe a cube in fancy-pants maths talk we say, 'A cube is a 3-dimensional shape with each face the same polygon, namely a square', and the same number of faces meeting at each vertex (a fancy word for corner).

That makes it a 'Platonic solid'. There are 5 Platonic solids altogether. The cube has a square face.

When we make a solid only out of triangles with all 3 sides equal (called an equilateral triangle) we get another 3 Platonic solids.

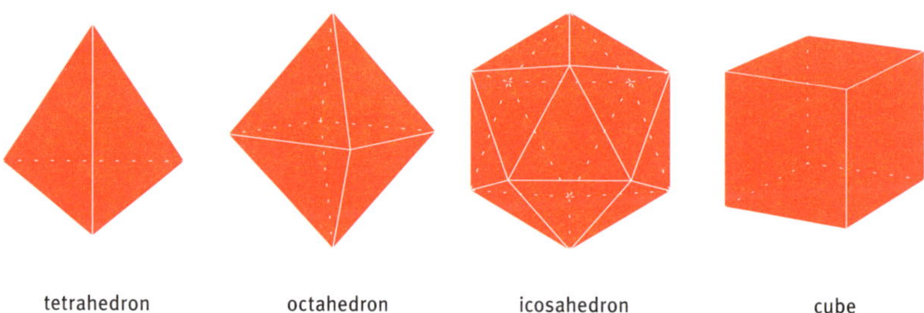

tetrahedron octahedron icosahedron cube

And when the faces are all regular pentagons (5 equal sides) we get ...

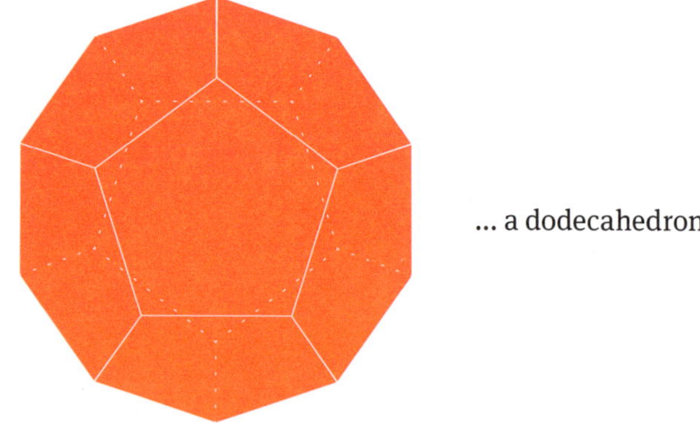

... a dodecahedron.

Five is the length of the hypotenuse (longest side) of the most famous right-angled triangle:

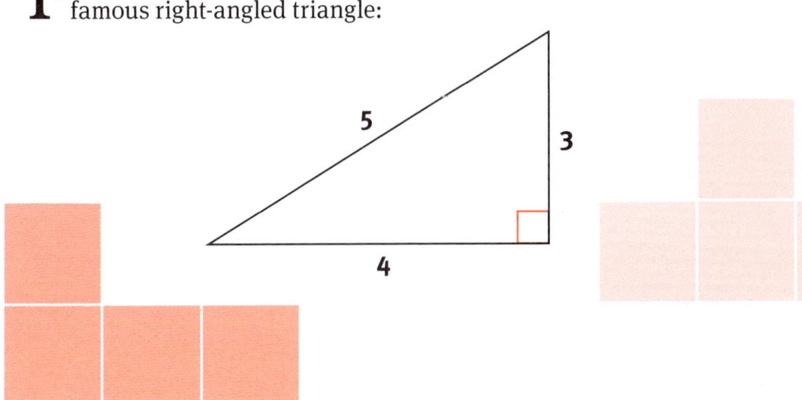

To square any number that ends in 5, just think of it as the number $n5$ where n is all the digits before the 5. To get your answer, write 25 and in front of it the number $n \times (n + 1)$. Huh? What is 75^2?

Well, here the $n = 7$ so $n \times (n + 1) = 7 \times 8 = 56$, so $75^2 = 5625$.

TETRIS

We call an arrangement of 4 squares all touching along their edges a 'tetronimo' (like a domino but with 'tetra' meaning 4).

If like me as a kid you didn't mind blowing a few bucks playing Tetris, you probably didn't realise that you were arranging the 5 basic or free tetrominoes.

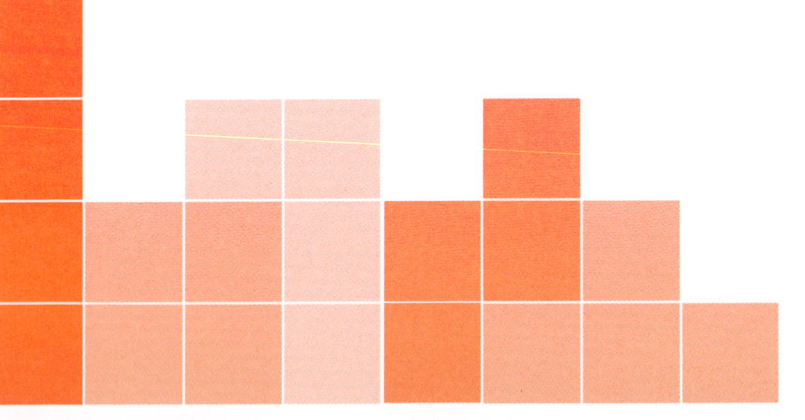

9

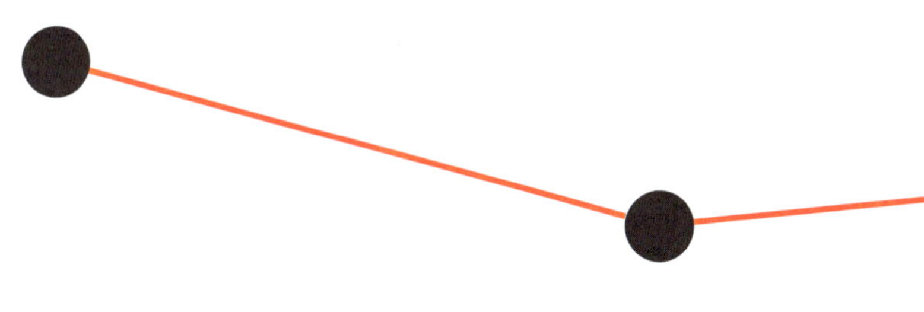

6 DEGREES OF KEVIN BACON

The phenomenon '6 Degrees of Kevin Bacon', inspired by the film *Six Degrees of Separation*, scores actors on how close they are to the famous ensemble actor via other co-stars.

For example, Russell Crowe was in *Cinderella Man* with Rance Howard who was in *Apollo 13* with Kevin Bacon, giving Russell Crowe a 'Bacon number' of 2.

This is an example of what mathematicians call a 'network' or 'path' in the world of movie actors.

It's extremely rare to find an actor with a Bacon number of 4 or higher, leading people to claim that Kevin Bacon is the centre of the Hollywood Universe. But once computer servers took over and crunched massive databases of film credits and the like it turns out Kevin Bacon is popular, but nowhere near the centre of the Hollywood Universe.

At the time of writing, the average Bacon number for any actor was 2.998, making our Kevin the 370th most connected actor in history. Winona Ryder (2.994, 342nd) and Whoopi Goldberg (2.941, 104th) are both more connected than Big Kev, as are the 3 most connected actors at the centre of the Hollywood Universe: Robert De Niro, Dennis Hopper and ... drum roll, please ... Harvey Keitel, with an average Keitel number of 2.849.

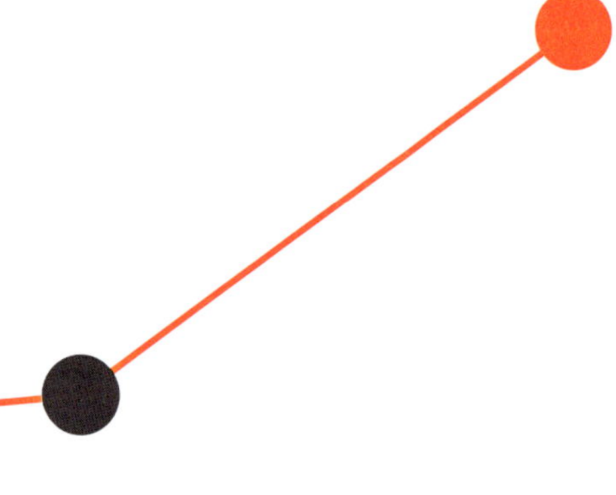

THE NUMBER OF THE BEAST

While mathematicians see the number 6 as perfect, some religiously inclined people get a bit freaked out when they see 6 ... or more specifically three 6s hanging out together.

In the Catholic Bible's 'Book of Revelations', the verse (Rev 13:18) reads, 'Let him that hath understanding hath the number of the beast: for it is the number of a man; and his number is six hundred threescore and six'.

A 'score' is 20, so that makes 666 the number of the beast.

I have no idea what that means and, to be honest, a lot of people a lot smarter than me have thought about it and can't agree what it means either. But I do know two things.

Firstly, the number of the beast can be written in many cool ways that involve 6s, for example:

$666 = 1^3 + 2^3 + 3^3 + 4^3 + 5^3 + 6^3 + 5^3 + 4^3 + 3^3 + 2^3 + 1^3$

$666 = 3^6 - 2^6 + 1^6$

$666 = 6 + 6 + 6 + 6^3 + 6^3 + 6^3$

And secondly, the Iron Maiden album *The Number of the Beast* (1982) absolutely rocks – especially the track 'Run to the Hills'.

FACTS ABOUT

Six is a 'perfect' number. Perfect numbers are numbers that are equal to the sum of their factors (not including themselves). So 6 can be written as 1×6 and 2×3. When we add the factors, $1 + 2 + 3 = 6$.

In addition to 6 equalling $1 + 2 + 3$, we also have $6 = 1 \times 2 \times 3$. To save time we write $6 = 3!$ which we say as '6 equals 3 factorial'. You might not think we've saved much time, but if you need convincing, try writing 100! out in full.

The 6-sided regular figure (meaning it has all sides equal in length) is called a hexagon. Hexagons are important building blocks in nature and can be found in beehives, snowflakes, tortoise shells and the structure of countless chemical compounds.

The only country with a *minimum* legal drinking age for consuming alcohol at home is the United Kingdom which prohibits such drinking below the age of 6 years. Fair enough, really.

While an adult male at rest has a heartbeat of around 72 bpm, a blue whale's heart beats just 6 times a minute. At the other end of the spectrum, it's estimated a canary's heart can beat anything up to 1000 times a minute – if they've really been working out.

Lastly, the amazing human small intestine is 6 metres long and full of all sorts of ... nah, you don't want to know.

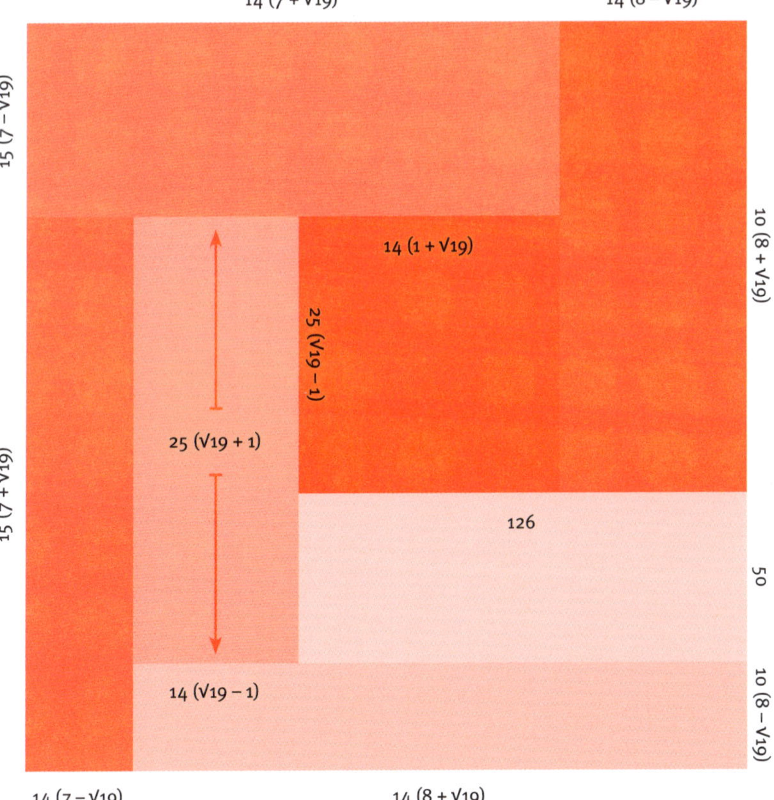

BLANCHE'S DISSECTION

Blanche's Dissection refers to the simplest dissection of a square into rectangles that all have the same area but are of different sizes.

There are 7 rectangles in Blanche's Dissection – the square is 210 units on each side – so its area is $210 \times 210 = 44{,}100$ square units. This means that each rectangle has an area of $44{,}100 \div 7 = 6300$ square units.

For another really cool way to split up a square, this time into smaller squares, see chapter 21.

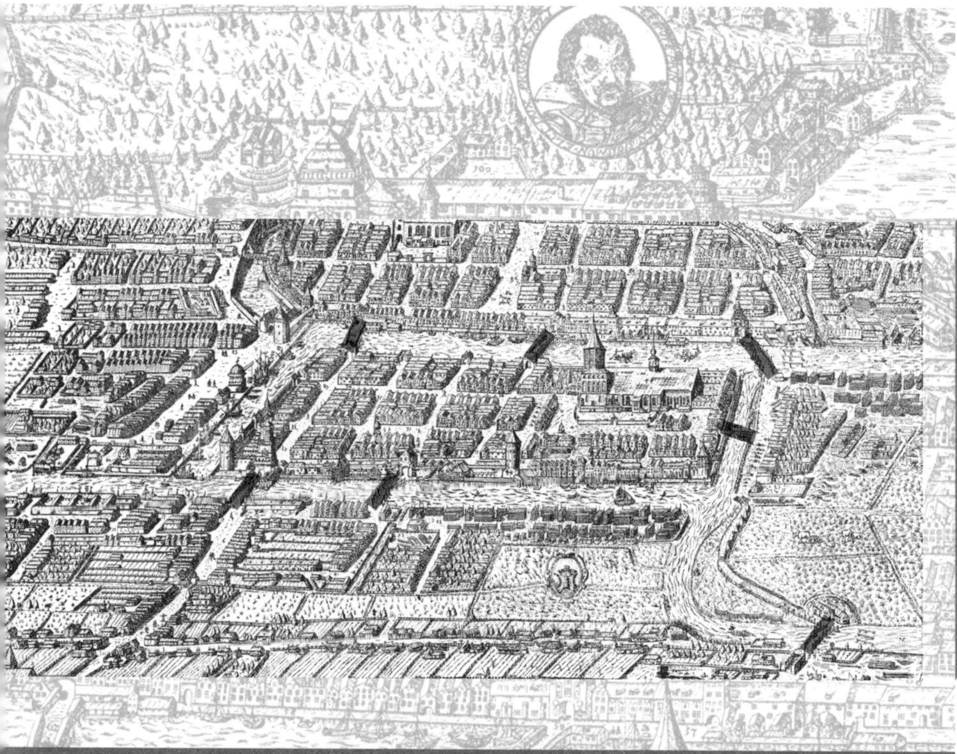

'THE BRIDGES OF KÖNIGSBERG' IS A FAMOUS OLD PROBLEM THAT DATES BACK TO THE 1730S. PUT SIMPLY, IT ASKS WHETHER OR NOT IT IS POSSIBLE TO TAKE A WALK THAT CROSSES ALL 7 OF THE OLD EUROPEAN CITY'S BRIDGES ONLY ONCE ...

If you've struggled to do it, don't panic, it can't be done! In 1736 Leonhard Euler proved that such a walk was impossible. This was far more than a discovery that helped out tourists in Königsberg. The entire area of mathematics known as 'graph theory' is said to have begun with Euler's solving of 'The Bridges of Königsberg' problem.

A recent survey of over 40,000 people by *The Guardian* website confirmed an often found result that when you ask people for their favourite number, they are most likely to say 7.

Between 1998 and 2006 Layne Beachley won seven – count 'em – *seven* surfing world championships. Awesome.

The opposite faces of a standard die add up to 7. The pattern of dots on the '5' side is is called a 'quincunx' – a fact I love, but just couldn't squeeze into chapter 5.

SEVEN CIRCLES

Draw a big circle and inside that, draw 6 smaller circles of whatever size you'd like that are all tangent to the big circle and their neighbours.

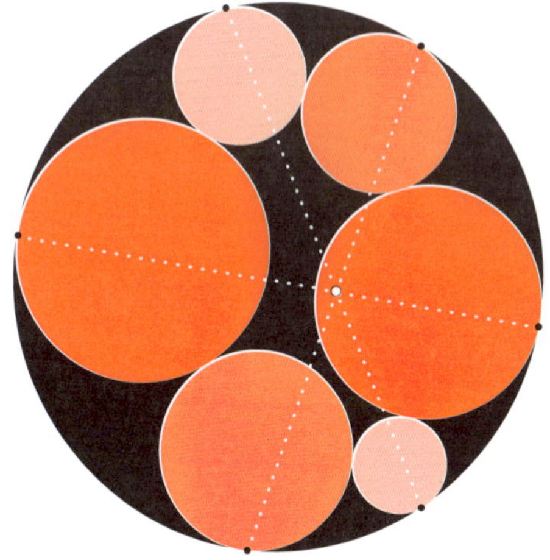

When you join up the tangent points of the opposite circles those 3 lines will always pass through a common point. Wow!

THAT SOMA-ZING!

The Soma Cube is a famous mathematical object – a cube made up of 7 pieces. There are actually 240 ways of combining these 7 pieces together.

8

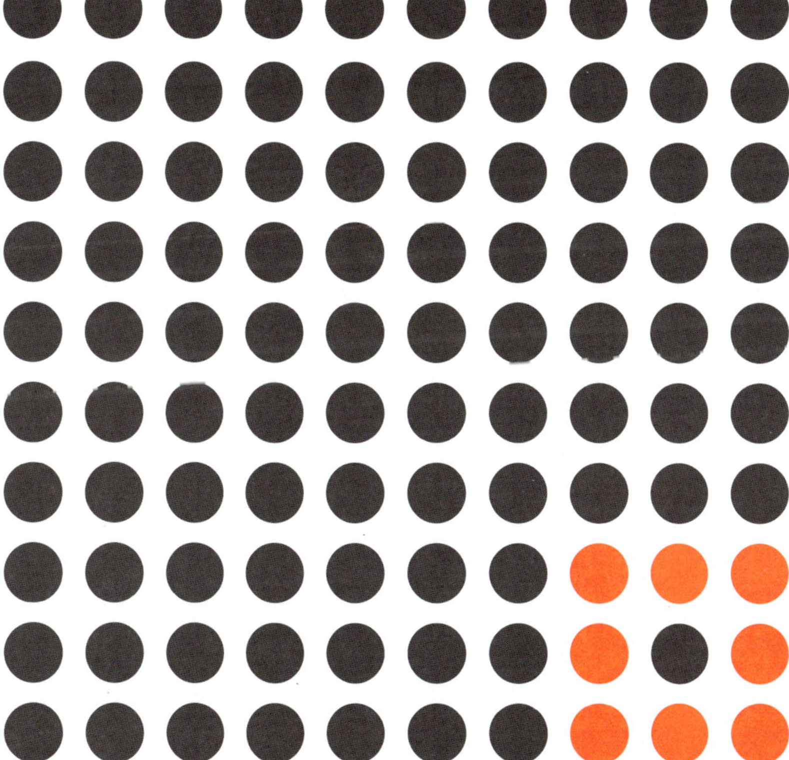

LIGHT YEARS

Though it takes only 8 minutes for a photon of light to reach Earth from the Sun, it took 100,000 years for that photon to come from the Sun's core to its surface!

$8 = 2^3$ & $9 = 3^2$

This is the only case of 2 consecutive numbers being powers of whole numbers – the fact that they both use 2 and 3 is just a cherry on top.

HIP TO BE SQUARE

Ancient Greek mathematicians discovered that when you square every odd number above 1, you get a number that is a multiple of 8 + 1. For example, $5^2 = 25$, which is $3 \times 8 + 1$.

OCTOBER – THE 8TH MONTH

It often puzzles people that October is actually the 10th month of the year – what with 'oct' meaning 8 and all (octagon, octopus and so on). The confusion usually lasts until the person is lucky enough to meet an expert in Roman history, in particular the Roman calendar.

Now to be honest, experts in Roman history can be a little thin on the ground these days, but this is where you come in. Next time someone wonders aloud, 'Why is October the 10th month of the year and not the 8th?' you can reply quick as a flash ...

'In the ancient Roman calendar, October was actually the 8th month. But when the Romans introduced names for 2 new months for the period that until then had simply been ignored because it was not productive farming time, October got bumped back to month number 10.'

When your friend replies, 'Wow, it's pretty impressive that you knew that' ... get used to it. This is going to happen to you a lot once you've finished reading this book.

8

Eight is a Fibonacci number. The Italian mathematician Fibonacci, or Leonardo of Pisa, wrote a famous book called *Liber Abaci* in 1202, which included the 'rabbit problem'. If 2 rabbits give birth to a new pair of baby rabbits each month, but the new pair don't breed until they're a month old, and the rabbits never die, how many pairs are alive each month?

The answer is:

1

1

2

3

5

8

13 ...

— The Fibonacci numbers!

EIGHT US PRESIDENTS HAVE DIED IN OFFICE. Of that number, 4 were assassinated (Abraham Lincoln, James A. Garfield, William McKinley, and John F. Kennedy) while 4 died of natural causes (William Henry Harrison, Zachary Taylor, Warren G. Harding and Franklin Delano Roosevelt). Coincidentally, there are a number of eerie similarities between the first president assassinated (Lincoln) and the last president assassinated (Kennedy). For instance, Lincoln was shot by John Wilkes Booth at Ford's Theatre; while Kennedy was shot by Lee Harvey Oswald in a Lincoln limousine, made by Ford ...

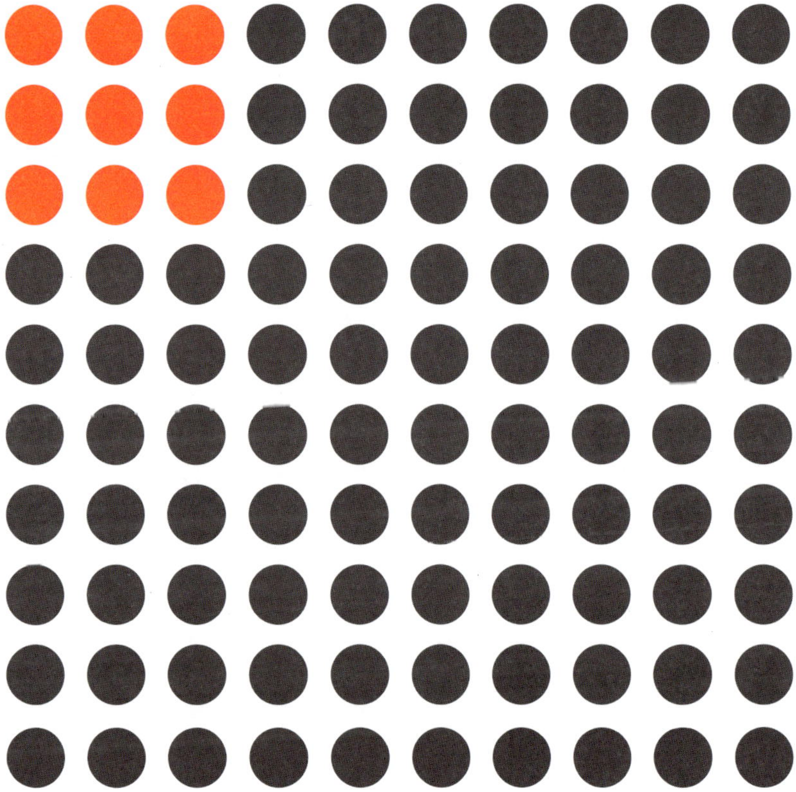

NINE-POINT CIRCLE

Here's a cool way to generate a '9-point circle' from a triangle. You can start with any triangle, but one like this with no angle bigger than 90 degrees (we call it an acute triangle) is easiest to use. First mark the 3 points that are mid-points of each side.

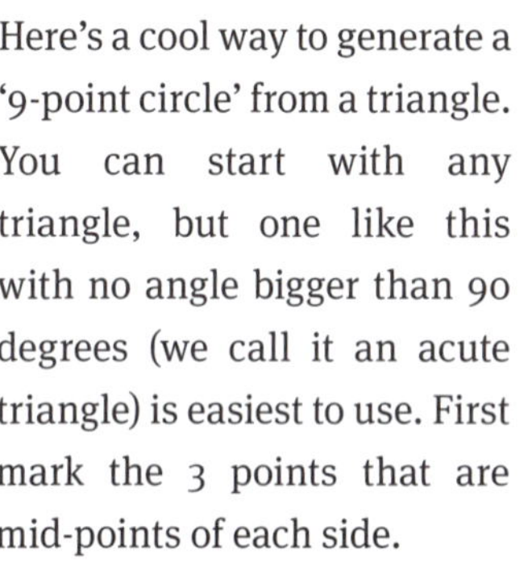

Then drop a line from each corner to hit the opposite side at 90 degrees (we call this an 'altitude'). This gives us another 3 points, and we notice that these altitudes all intersect at a common point.

Now mark the point halfway between the common point and each corner (by the way, we call the corner a 'vertex') giving us a final 3 points.

You'll never guess what we can draw through those 9 points – well, in fact, you probably will guess and if you can't, look back to where I mentioned a '9-point circle'. That's right, cool huh?

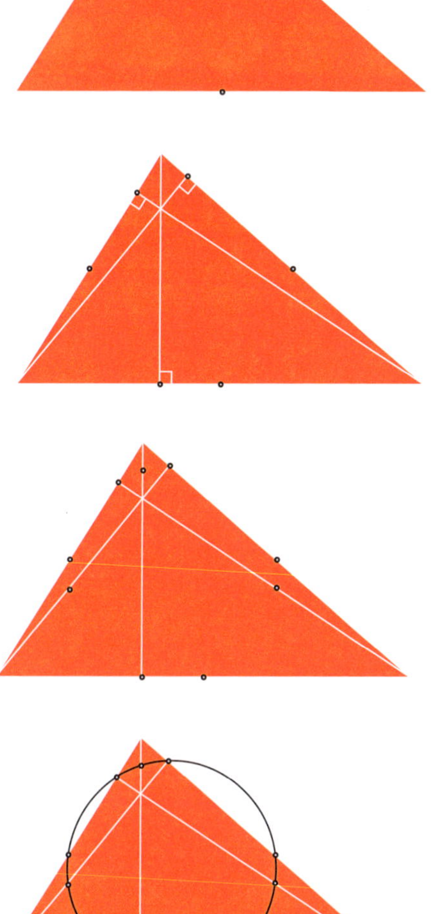

ALPHAMETICS

An alphametic is a puzzle which is written in words, but for which you change the letters of the words to numbers to try to make an equation that works out.

A famous example, from 1924, is this:

SEND +
MORE
———
MONEY

Where the letters of the words correspond to the numbers 0–9.

To solve this puzzle, I'll give you a couple of hints, then you're on your own.

First up, if we let S and M be as big as possible, we'd still only be adding '9 thousand and something' to '8 thousand and something' so there is no way the 5-digit answer could be 20,000 or higher.

So M must be equal to 1.

But there is also an M at the beginning of the second line, so it is also a 1.

Now we have:

SEND +
1ORE
———
1ONEY

So the S can't be 7 or less – if it was, the first line wouldn't be big enough to push us over the 10,000 mark for the answer.

So S = 8 or S = 9.

But whether S = 8 or 9, the first line will only be '8 thousand and something' or '9 thousand and something'. Either way, we can't make the first line large enough to get an answer of 12,000 or more for the equation. So O = 0 or 1. But we already have M = 1, so O = 0.

This gives us:

SEND +
10RE
———
10NEY

We can look at the column with E, 0 and N in it and notice this – if we are not carrying a 1 from the previous column, then we'd get E + 0 = N in this column. And that wouldn't make sense, because E and N have to be different numbers. So we must have to carry a 1 from that second column, and we must have that E + 1 = N.

Now let's try and work out if S = 8 or 9.

If S = 8, we still have to get an answer over 10,000. So E = 9 is our only choice. But that doesn't work, because we can't then get E + 1 = N (here N would be 0 and we'd carry another 1 across to the next column. But the 0 is already taken).

So S = 9.

This sort of stuff isn't easy first time. So go through it all again – writing it down with a pen and paper – and make sure you understand it. When you do, you'll see that so far we have:

```
  9END +
  10RE
 ______
 10NEY
```

And we know E + 1 = N.

The numbers we have left are 2, 3, 4, 5, 6, 7 and 8 and we must allocate five of them to the letters E, N, D, R, Y.

Well E = 8 would give N = 9 and that is a no-no.

Give it a go yourself from here. The two things to do are 1) just try different options for E and N satisfying E + 1 = N and 2) notice something very important about the value of R.

You'll find the answer at the back of the book.

MAGIC SQUARE

The numbers 1 to 9 can be arranged in a 3 × 3 grid such that every row, column and diagonal add up to the same number – in this case 15. We call such an arrangement a 'magic square'. I hope you like it, because we'll meet quite a few of them in this book.

2	7	6
9	5	1
4	3	8

QUIZ QUESTION: **Of what possible mathematical use is the 10-word phrase, 'Can I have a large container of coffee? Thank you'?** HINT: Count the number of letters in each word ... ANSWER AT THE BACK OF THE BOOK

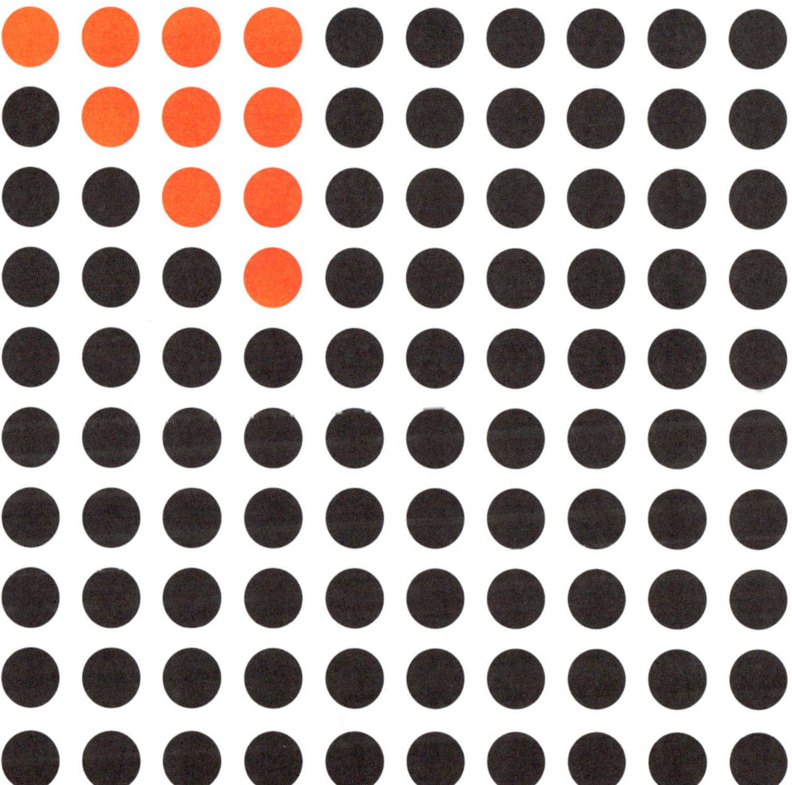

10

Why is 10 scared of 7? Because 7, 8, 9 ... and 10 is next. (Sorry, dad joke.)

BASE-10

Our number system is called 'base-10'. This means that after counting 0, 1, 2, 3 ... 9 we move across to the next 'place' and count 10, 11, 12, 13 and so on. The first place of a 2-digit number tells you how many times we've already counted to 10, so 35 means 3 lots of 10 and 5 lots of 1. If we used, say, base-7, then the number 35 would be $3 \times 7 + 5 \times 1$, which is 26 in our base-10. When we have filled up the second place column, we start a third place. In base-6, 312 would be $3 \times 6 \times 6 + 1 \times 6 + 2$ which is 116 in base-10.

TEN CHESS MOVES

According to David Darling's excellent *Universal Book of Mathematics* there are 169,518,829,100,544,000,000,000,000,000 different ways of playing the first 10 moves of a chess game.

But it should be said ... a lot of those possible moves, while legal, would be pretty average.

FACTORIAL FUN

In chapter 24 you'll meet what mathematicians call a 'factorial' and learn that $4! = 4 \times 3 \times 2 \times 1 = 24$.

Well $10! = 6! \times 7!$ (you can prove this yourself if you're game – with pen and paper if you really back yourself).

This is the only case of two consecutive factorials multiplying to give us a different one.

The great Indian mathematician Srinivasa Ramanujan, who we'll learn more about in chapter 54, recognised that $9^3 + 10^3 = 1^3 + 12^3$.

GOSPER'S 10 ISOSCELES TRIANGLES

American mathematician William Gosper is a fascinating guy. Some people consider him to have founded the computer hacker community and at one time he held the world record for calculating π to a previously unmatched 17 million decimal places.

He also created this beautiful dissection (above) of a square into 10 isosceles triangles (triangles with 2 equal sides).

Excepting the obvious case of simple diagonals across the square, this can't be done with less than 10 triangles. Well played, Bill.

Decapoda are crustaceans which have 10 legs – hermit crabs, for example.

Check this out: $1^2 = 1$, $11^2 = 121$, $111^2 = 12{,}321$ and $1111^2 = 1{,}234{,}321$ … can you see where this is going? Next time you really want to impress someone, bust out a bit of $111{,}111{,}111^2 = 12{,}345{,}678{,}987{,}654{,}321$.

ELEVEN SQUARES

The 11 squares in the diagram below together don't form a square but go awfully close. Each number indicates the side length of each square. In fact, they form a rectangle which is 176 × 177.

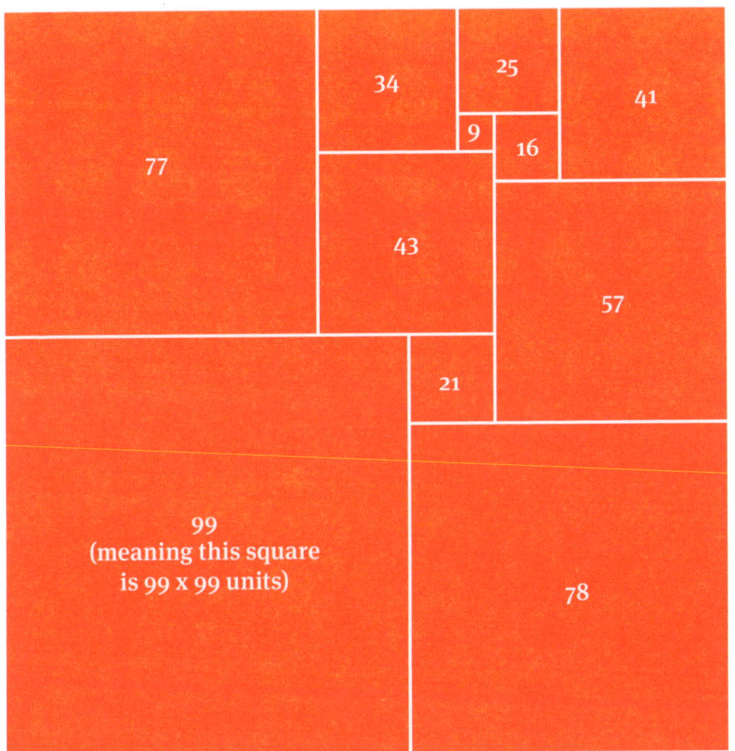

Later on (check out chapter 21) we'll see that we need a few more squares before we can fit them together to form a bigger square.

HEXOMINOES

There are 35 free hexominoes (see also tetrominoes in chapter 5) altogether. But of the 35, what do the following 11 have in common?

Answer: They can all be folded into a cube.

REP-RESENT

The number that is just 1 repeated *n* times is written as R_n and we call it a 'rep-unit', an abbreviation of 'repeated unit'; for example, $R_2 = 11$, $R_6 = 111,111$.

We know that if *n* divides *m*, R_n divides R_m and that apart from $R_1 = 1$, no R_n is a square, a cube or any higher power.

A rep-unit that is also a prime number is called a 'rep-unit prime'. Clearly $R_2 = 11$ is prime.

It takes a while to find another rep-unit prime, but

$R_{19} = 1,111,111,111,111,111,111$ is prime

As are R_{23}, R_{317} and R_{1031}

If the idea of R_{1031} blows your mind, get this – $R_{270,343}$ is said to be probably prime. Which means exactly what it sounds like. We *think* it's prime ... but it's just way too big for us to have cracked it yet!

QUIZ QUESTION: Show that the rep-units R_3, R_4 and R_6 are not prime by factorising them. ANSWER AT THE BACK OF THE BOOK

MULTIPLICATIVE PERSISTENCE

If I take a number and multiply all of its digits, I will get a new number. For example, if I start with 748, I get $7 \times 4 \times 8 = 224$. I could keep doing this and I'd go from 748 to 224, to $2 \times 2 \times 4 = 16$ to $1 \times 6 = 6$ and then I'd be stuck at 6.

I can do this with any number, and eventually I get down to a 1-digit number and have to stop.

For the example of 748, it took me 3 steps to reach the end of the chain, so we say that 748 has a 'multiplicative persistence' of 3.

It seems that no matter how big the number, they all have a multiplicative persistence of 11 or less.

This is one of those wonderful mathematical concepts called a 'conjecture' ... we've tested it up to the ridiculously large number 1 followed by 233 zeros and not found anything to prove us wrong – but that's not the same as proving it for *all* numbers.

The smallest number with multiplicative persistence of 11 is 277,777,788,888,899.

QUIZ QUESTION: Prove that the multiplicative persistence of 277,777,788,888,899 is 11. ANSWER AT THE BACK OF THE BOOK

TURN IT UP

Eleven is the loudest the Marshall guitar amps go in the hilarious rock-umentary *This is Spinal Tap* (kids, trust me, if you haven't seen it, watch it *now*!). As Spinal Tap's guitarist Nigel Tufnel proudly explains, while all other bands' amps stop at 10, '... what we do is, if we need that extra push over the cliff [we go to] 11. One louder'.

When the 'interviewer' asks, 'Why don't you just make 10 the top number and make that a little louder?' Nigel waits for an awkward amount of time then deadpans ... 'These go to 11'. Gold.

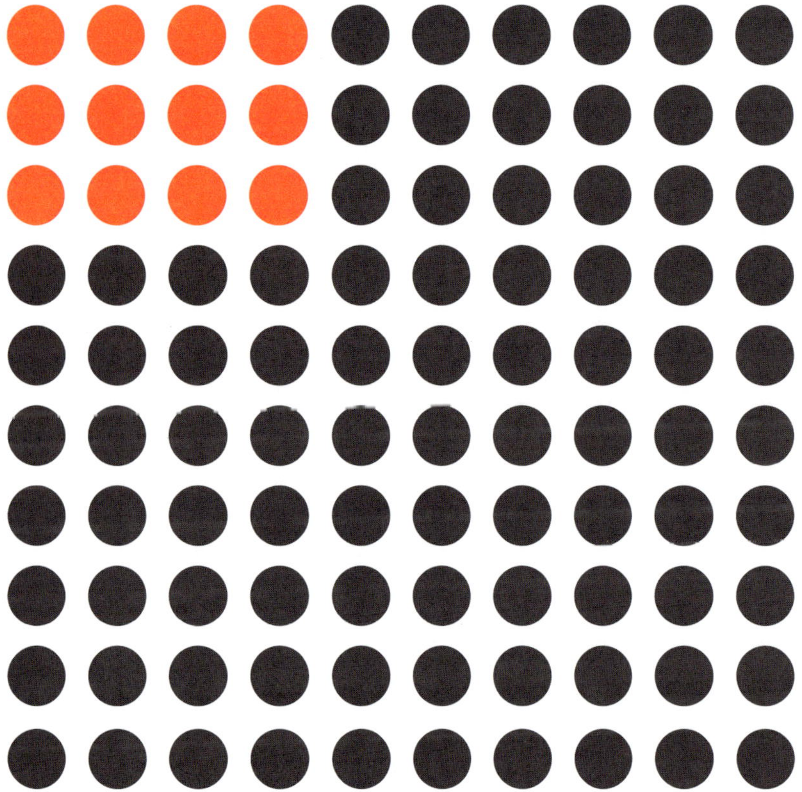

12

ABUNDANT!

12 is the first 'abundant' number. A number is abundant if the sum of all its factors apart from itself is greater than the number. So 12 can be written as 1×12, 2×6, or 3×4. Add up $1 + 2 + 3 + 4 + 6$. They equal 16, and 16 is bigger than 12, therefore 12 is abundant.

EXCLUSIVE!

Probably the most exclusive (and surely the coolest) club in history comprises only 12 men. They are the astronauts from the Apollo missions between 1969 and 1972 who are the only men to have walked on the Moon.

HILARIOUS!

Fawlty Towers, *The Young Ones* and *The Office* are all hilarious, all British and each had only 12 episodes ever made (excluding the two Christmas specials they made of *The Office*).

THE DOZENS

Twelve is also known as a 'dozen'. Dozen comes from the Latin word *duodecim*, *duo* meaning two, and *decem* meaning ten.

Throughout history, twelve has played an important part in many counting systems and we still see traces of that today – 12 hours from noon till midnight, 12 pence in the old shilling in the UK, and so on. Twelve has the benefit of dividing into half, thirds or quarters and giving a whole number answer, as opposed to 10.

Twelve also features in a very neat little piece of mathematical poetry (see the Quiz Question below) but, in the same way that 12 is a dozen, you might need to know the other name for some of the numbers: 144 is a *gross* and 20 is a *score*.

QUIZ QUESTION: Decipher the limerick below

$[(12 + 144 + 20) + 3\sqrt{4}] \div 7 + 5 \times 11 = 9^2 + 0$

ANSWER AT THE BACK OF THE BOOK

PENTOMINOES

When we join 5 squares together by their edges we get a 'pentomino'.

There are 12 different pentominoes, and here they all are:

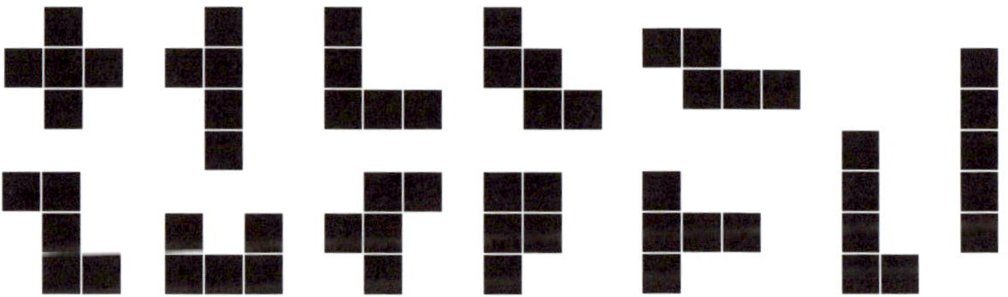

The 12 pentominoes each contain 5 squares so, all told, they contain $5 \times 12 = 60$ squares. They can fill a 5×12 grid like this:

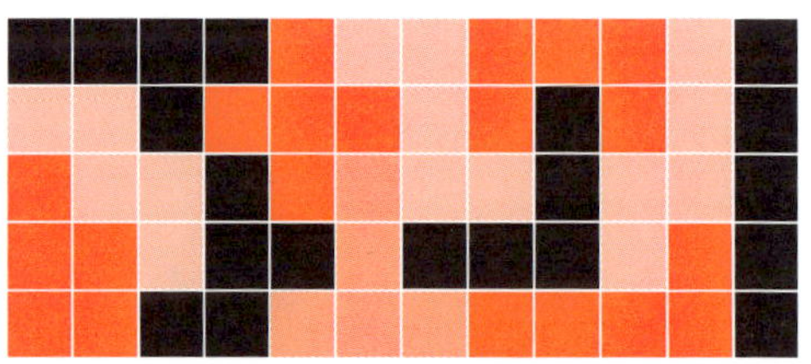

Actually, there are 1010 different ways to arrange them in a 5×12 grid. But 60 is also 3×20 or 6×10 or 4×15. There are only

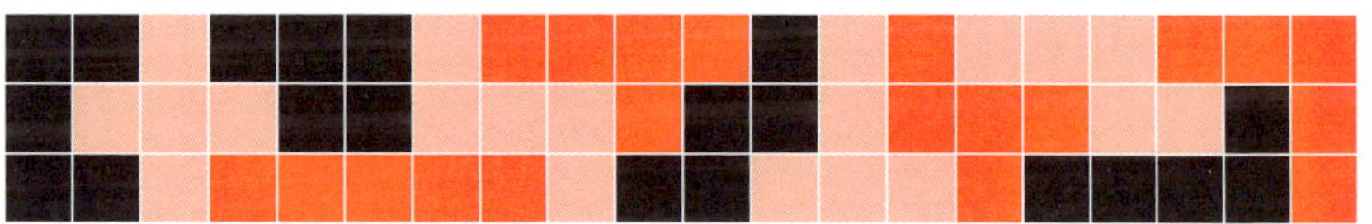

two ways of filling a 3×20 grid, but 368 ways of making a 4×15 grid and 2339 possible 6×10 rectangles out of the 12 pentominoes. Nice.

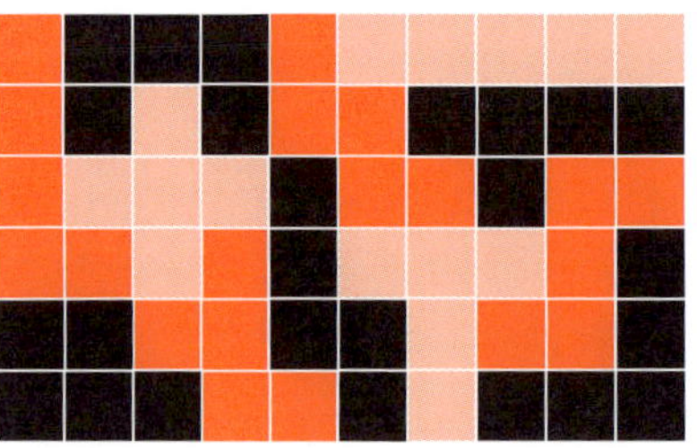

The middle section of your spine contains 12 thoracic vertebrae. You think that's impressive? Some sloths have 25!

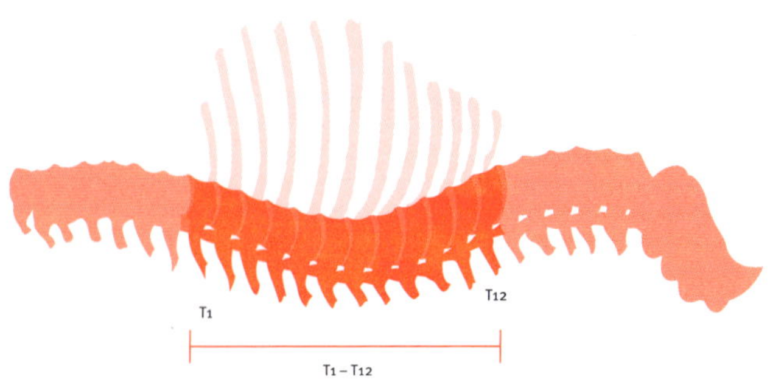

THE QUEENS PROBLEM

In chess a queen can attack in every direction:

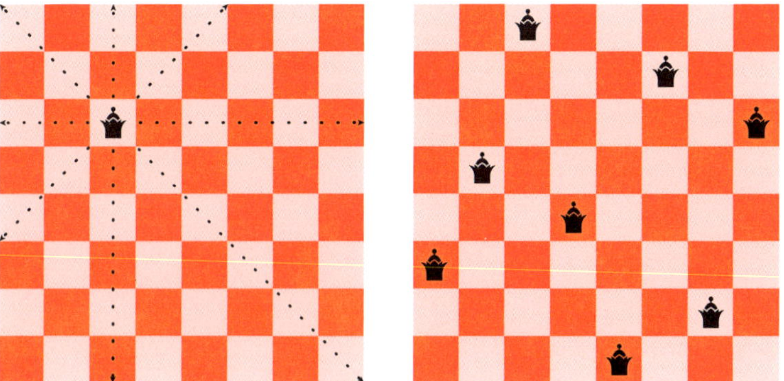

But it's possible to put 8 queens on an 8 × 8 chessboard in a way that none of them threaten each other.

In fact, there are 12 different ways of doing this.

QUIZ QUESTION: Find the 12 ways to place 8 queens on a chessboard so that no two attack each other. Note, these solutions are different even if you rotate them and reflect them.

ANSWER AT THE BACK OF THE BOOK

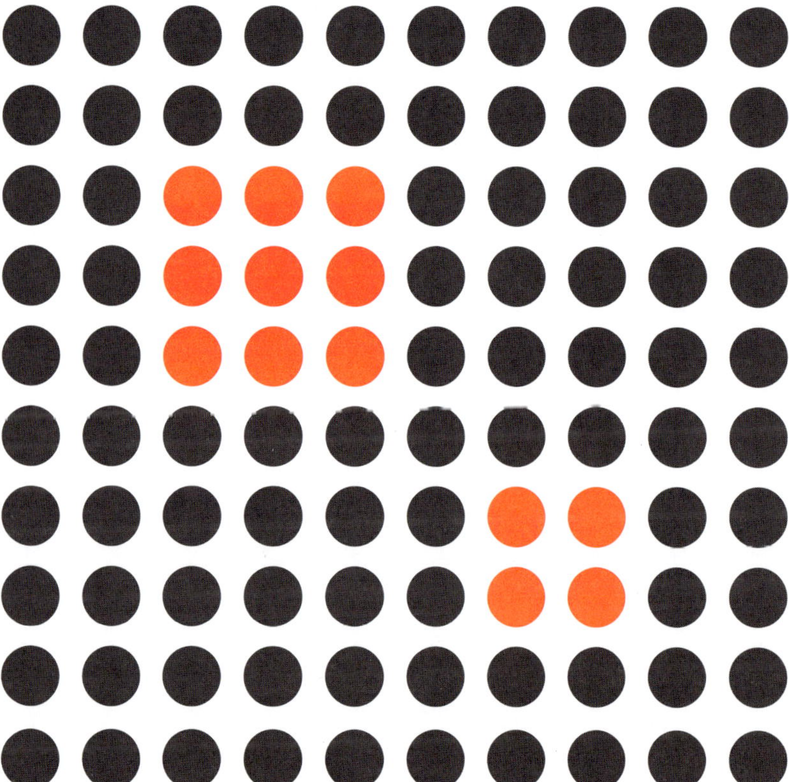

13

If you freaked out when you saw this chapter, you may suffer from *triskaidekaphobia* – fear of the number 13. We're so scared of the number 13 it costs the world billions of dollars a year in absenteeism, travel cancellations and loss of business. There were 13 people at the Last Supper, witches apparently hang out in covens of 13, and, in China, the 13th month that was occasionally added to bring the lunar year into line with the solar year was called 'Lord of Distress' or 'Oppression'. In most hotels you won't find room numbers 713 or 2613, for instance, because they contain the number 13. In fact, often the whole 13th floor is left out, just to be safe.

MAGIC SQUARES

Opposite is one of the most impressive and mysterious magic squares of all time (for more about magic squares see chapters 9, 34 and 64).

Starting with the 3 × 3 magic square in the centre, we can see that all rows and columns add up to a 'constant' of 16,311. This is 10,874 more than the 5437 at its centre.

But this 3 × 3 square sits 'embedded' in a 5 × 5 square, with a constant of 27,185, which is 10,874 more than our original constant. And this sits in a 7 × 7 square with a constant of 38,059, again an increase on the previous constant of 10,874. You'll never guess what happens with this 7 × 7 magic square. That's right, it sits inside a 9 × 9 square with constant of 38,059 + 10,874 = 48,933.

The 11 × 11 magic square you'd get by deleting the outside rows and columns of the original grid has a constant of 59,807 (again an increase of 10,874) while the final 13 × 13 magic square has a constant of ... that's right ... 59,807 + 10,874 = 70,681. Further, every entry in the entire 13 × 13 square is a prime number.

To add a final layer of mystique, this magic square was published in the *Journal of Recreational Mathematics* in October 1961 having been sent to 'Francis L. Miksa of Aurora, Illinois from an inmate in prison who, obviously, must remain nameless'.

Thirteen is a Fibonacci number. Fibonacci numbers regularly occur outside of mathematics. Almost all apples, roses, blackberries, raspberries, strawberries, peaches, plums, pears and cherries have flowers with 5 petals, and 5 is a Fibonacci number. Pineapples have either 8 or 13 spirals coming out of their top. These are called *parastichies*.

If it's Friday when you're reading this, and it's the 13th, you'll know this month started on a Sunday. Makes sense when you think about it ...

1153	8923	1093	9127	1327	9277	1063	9133	9661	1693	991	8887	8353
9967	8161	3253	2857	6823	2143	4447	8821	8713	8317	3001	3271	907
1831	8167	4093	7561	3631	3457	7573	3907	7411	3967	7333	2707	9043
9907	7687	7237	6367	4597	4723	6577	4513	4831	6451	3637	3187	967
1723	7753	2347	4603	5527	4993	5641	6073	4951	6271	8527	3121	9151
9421	2293	6763	4663	4657	9007	1861	5443	6217	6211	4111	8581	1453
2011	2683	6871	6547	5227	1873	5437	9001	5647	4327	4003	8191	8863
9403	8761	3877	4783	5851	5431	9013	1867	5023	6091	6997	2113	1471
1531	2137	7177	6673	5923	5881	5233	4801	5347	4201	3697	8737	9343
9643	2251	7027	4423	6277	6151	4297	6361	6043	4507	3847	8623	1231
1783	2311	3541	3313	7243	7417	3301	6967	3463	6907	6781	8563	9091
9787	7603	7621	8017	4051	8731	6427	2053	2161	2557	7873	2713	1087
2521	1951	9781	1747	9547	1597	9811	1741	1213	9181	9883	1987	9721

So 13 = 11 + 2 = 12 + 1, right? Yeah, whatever, Adam, you're grasping at straws here, pal. But look at this: when we write 13 = ELEVEN PLUS TWO = TWELVE PLUS ONE we see that the two word equations are anagrams of each other.

GOOOOOOOAAAAAAALLLLLLL!

We've met the Platonic solids (see chapter 5), which are 3-dimensional solids with every face the same regular shape, and the same number of faces meeting at each vertex.

If we look at solids with *more* than one type of regular face (and the same set of shapes meeting at each vertex), we get to the Archimedean solids.

There are 13 in total and you'll meet quite a few throughout this book. Probably the best known is the 'truncated icosahedron':

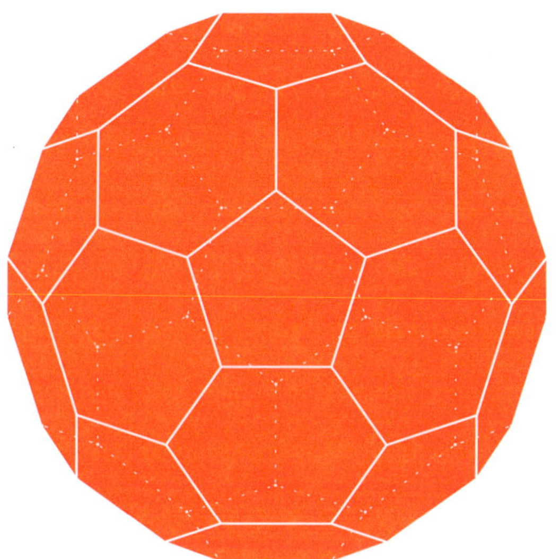

... an arrangement of 12 pentagons (5 sides) and 20 hexagons (6 sides) which most of us know as the 'soccer ball' shape because from the 1970s onwards it was the most popular design of footballs around the world.

カレ

I don't pretend to know about the 14 different families by which we classify fragrances – but I'm reliably informed they are floral; soft floral; floral oriental; soft oriental; oriental; woody oriental; woods; mossy woods; dry woods; aromatic; citrus; water; green and fruity.

UNWELCOME CLUBS

When playing PGA golf you can carry no more than 14 clubs in your bag. For most amateur hackers like me, having more than about 6 clubs tends to freak us out, but there have been occasions where professional golfers have been penalised for accidentally having too many clubs in their bag. At the 2001 British Open, the great Welsh golfer Ian Woosnam was in a potentially winning position on the final day when his caddie uttered the immortal words, 'You're going to go ballistic ... we've got *two* drivers in the bag.'. Woosnam was penalised 2 strokes for carrying 15 clubs. Understandably, he threw his extra driver into the woods!

CARBON-14

Carbon is an element crucial to life. On Earth, carbon occurs in 3 natural forms (or isotopes). All 3 of these have 6 protons, but Carbon-12 has 6 neutrons, Carbon-13 has 7 neutrons and Carbon-14 has 8 neutrons.

Carbon-12 makes up approximately 99% of the carbon on Earth and Carbon-13 makes up about 1%. That doesn't leave much room for Carbon-14, in fact only 0.0000000001% or 1 part per trillion of all carbon on Earth is Carbon-14.

Carbon-14 is radioactive, with a half-life of about 5730 years. Measuring how much of it is in stuff is the cornerstone of 'carbon dating', central to working out the age of fossils and so on, giving us a fascinating look back into history.

CATALAN NUMBERS

There are 5 ways of dividing a pentagon (a 5-sided figure) into 3 triangles:

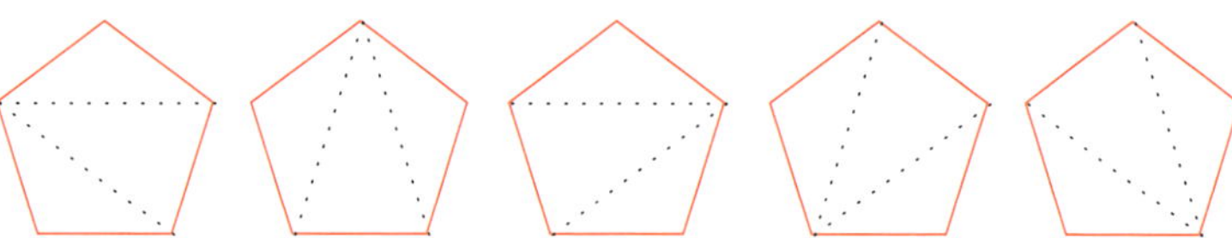

The formula for the number of ways you can divide a regular polygon with $(n + 2)$ sides into n triangles is given by the equation:

$$C(n) = \frac{(2n)!}{n!(n+1)!}$$

... which I know is enough to make you almost lose your lunch, but trust me isn't that bad. Remember that $3! = 3 \times 2 \times 1 = 6$ and $4! = 4 \times 3 \times 2 \times 1 = 24$.

So to divide a pentagon into 3 triangles, let $n = 3$ and the Catalan equation simply becomes:

$$C(3) = \frac{6!}{3! \times 4!}$$

$$= \frac{6 \times 5 \times 4 \times 3 \times 2 \times 1}{(3 \times 2 \times 1) \times (4 \times 3 \times 2 \times 1)}$$

... and the $4 \times 3 \times 2 \times 1$ cancel top and bottom leaving us with:

$$C(3) = \frac{6 \times 5}{3 \times 2 \times 1} = 5$$

So there are 5 ways to divide a pentagon into 3 triangles.

If you're really feeling game, convince yourself that $C(4) = 14$ and that there are 14 ways to divide a hexagon into 4 triangles ... here are 2:

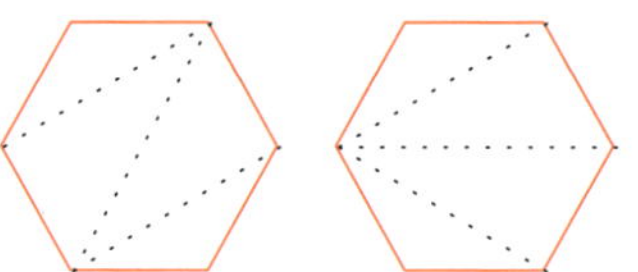

THE TRUNCATED CUBE

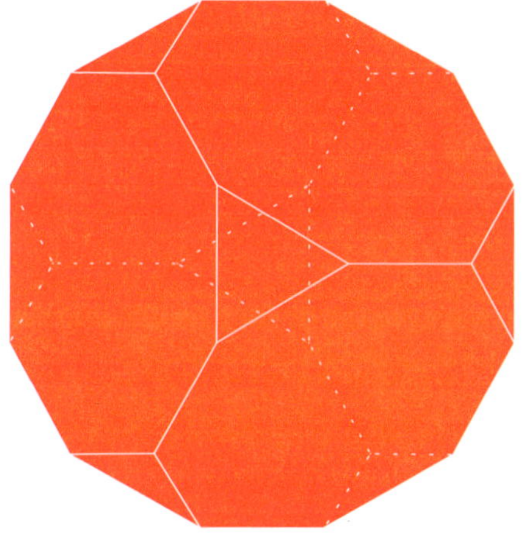

A 'truncated' cube has 14 faces, 24 vertices and 36 edges. Check that Euler's formula $V + F - E = 2$ (see chapter 3) applies to this Archimedean solid.

THE STOMACHION

The ancient puzzle known as the 'stomachion' contains 14 pieces which can be arranged in a square.

Though around since at least the time of Archimedes (287–212 BC), it wasn't until 2003 that Bill Cutler used a computer to show that this can be done an amazing 17,152 ways. If you remove solutions that are rotations and reflections of each other, Cutler found 536 unique solutions.

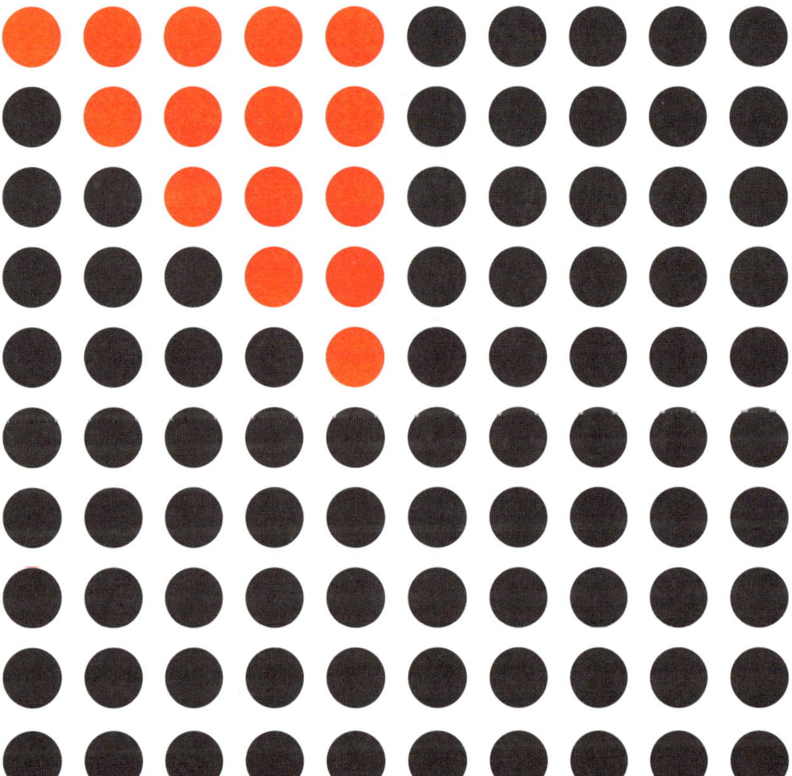

15

THE FAMOUS FIFTEEN PUZZLE

The geek puzzle that defined the 1980s was the Rubik's Cube (see chapters 20 and 43). For a few months in the 1880s *this* was all the rage:

2	1	12	11
13	9	15	10
3	14	8	
4	5	6	7

What is it, you ask? It's Noyes Chapman's famous 'Fifteen Puzzle' where you have to slide squares into the open space and rearrange them to eventually get them reading, from top left, 1 – 2 – 3 ... – 14 – 15.

If you take the tiles out and replace them randomly, the puzzle will only be solvable 50% of the time. For example:

1	2	3	4
5	6	7	8
9	10	11	12
13	15	14	

... is *unsolvable.*

In a beautiful connection between geeky games, the chess supremo Bobby Fischer could solve a Fifteen Puzzle in 20-odd seconds. Show-off!

LEGEND

Australian basketball champ Lauren Jackson wears the number 15 in honour of her mother, Maree, who wore the same number at basketball world championships in the 1970s.

POOL PARTY

Because 15 is the fifth triangular number it is the number of coloured balls in a standard game of pool.

UNCOPYRIGHTABLE

With 15 different letters, 'uncopyrightable' is the longest word in English with no repeated letters.

It actually shares the title with the considerably more obscure 'dermatoglyphics', the study of fingerprints.

While the standard champagne bottle holds 750 millilitres, there are actually 15 different sizes of bottle from which you can choose, ranging from the ¼ bottle piccolo to the massive 30-litre Midas or Melchizedek, which is named after the Biblical King of Salem. I can only presume he was quite the party animal.

Piccolo (0.25 bottle)
Demi (0.5 bottle)
Standard (1 bottle)
Magnum (2 bottles)
Jeroboam (4 bottles)
Rehoboam (6 bottles)
Methuselah (8 bottles)
Salmanazar (12 bottles)
Balthazar (16 bottles)
Nebuchadnezzar (20 bottles)
Melchior (24 bottles)
Solomon (26.66 bottles)
Sovereign (33.33 bottles)
Primat **or** Goliath (36 bottles)
Melchizedek **or** Midas (40 bottles)

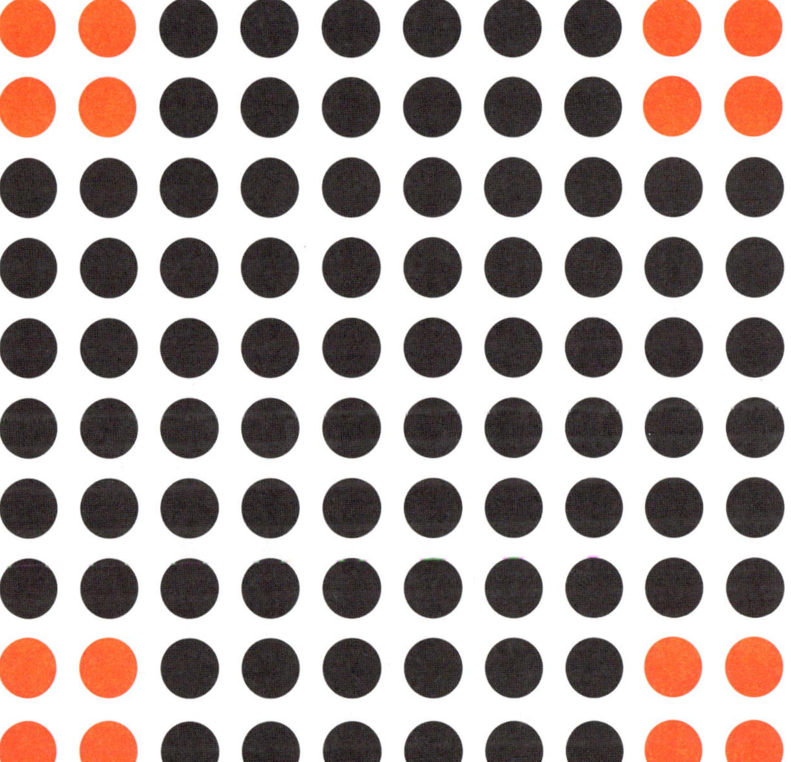

16

SWEET SIXTEEN

Most books are printed using massive machinery the size of your average semi-trailer. These machines, known as 'offset' printers, work fast and furious, cutting huge rolls of paper down into what are known as 'sections'. Each section is then divided up into the number of pages required. The most economical number of pages per section is 32, though sometimes there can be 16 pages per section, or even 8 pages.

But before you start bending this little baby back at the spine to check on the number of pages per section, I can confirm it has been printed in sweet sections of 16.

OUNCES

Those of us who measure in pounds and ounces would know that there are 16 ounces in a pound. But hold your horses, it's not that simple.

In Roman times the pound was 12 ounces and that remains to this day as the 'Troy ounce', which is used to measure gold, gemstones and so on.

But everything else is measured in *Avoirdupois* pounds (from the French *avoir de pois* meaning 'goods of weight') which is 16 ounces. The advantage of 16 ounces to the pound is that 16 can easily be halved and halved again and again *and again* before you get to 1 ounce.

Our old pal Leonhard Euler proved that 16 is the only number that can be written in a reverse power way: $16 = 2^4 = 4^2$.

RELATIVELY PRIME

The numbers 5 and 6 are called 'relatively prime' because they do not have a common factor.

If we look at the numbers 4, 5 and 6, we see that 4 and 6 are *not* relatively prime because 2 divides into both of them, but 5 *is* relatively prime to both of them. So this list of consecutive integers contains at least one number (5) that is relatively prime to all the others on the list.

Here's something cool.

Any list of 16 or fewer consecutive numbers *must* have at least one number on it that is relatively prime to all others on the list.

The first list of 17 numbers that fails this test is the one from 2184 up to 2200.

If you go further into the counting numbers you'll find a list of 18 consecutive numbers that have no number relatively prime to all the others, and further on again a list of 19 numbers; in fact, for any number n you choose, we can eventually find a list of n consecutive numbers with no member relatively prime to all the others.

Just another example of the hidden complexity of prime numbers.

SQUARE CIRCLES

Sixteen circles can be arranged in a square:

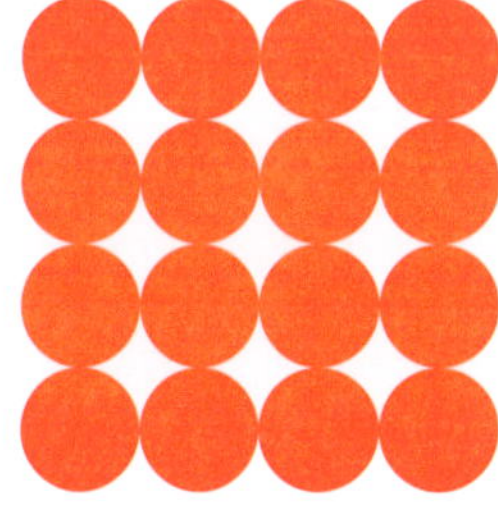

But they can also become the outline of a square:

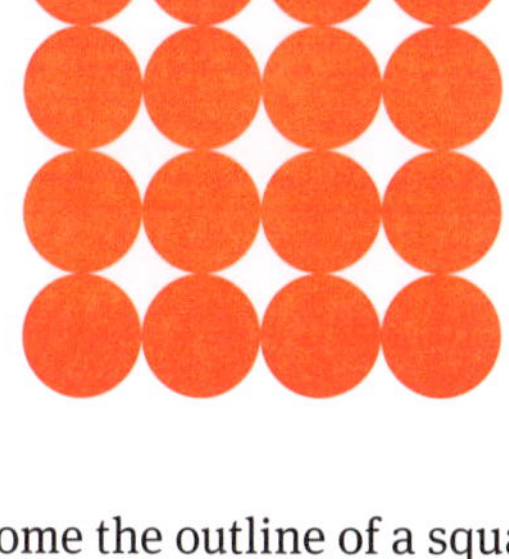

What the first diagram shows is that $16 = 4 \times 4$ or 4^2.

And the second shows us that a 5×5 square with a 3×3 square removed gives us the remaining outline, or $5^2 - 3^2 = 4^2$.

But this only happens for the consecutive numbers 3, 4 and 5, not for any other example of 3 consecutive counting numbers.

So 16 is the only number of circles for which you can make two related squares like this.

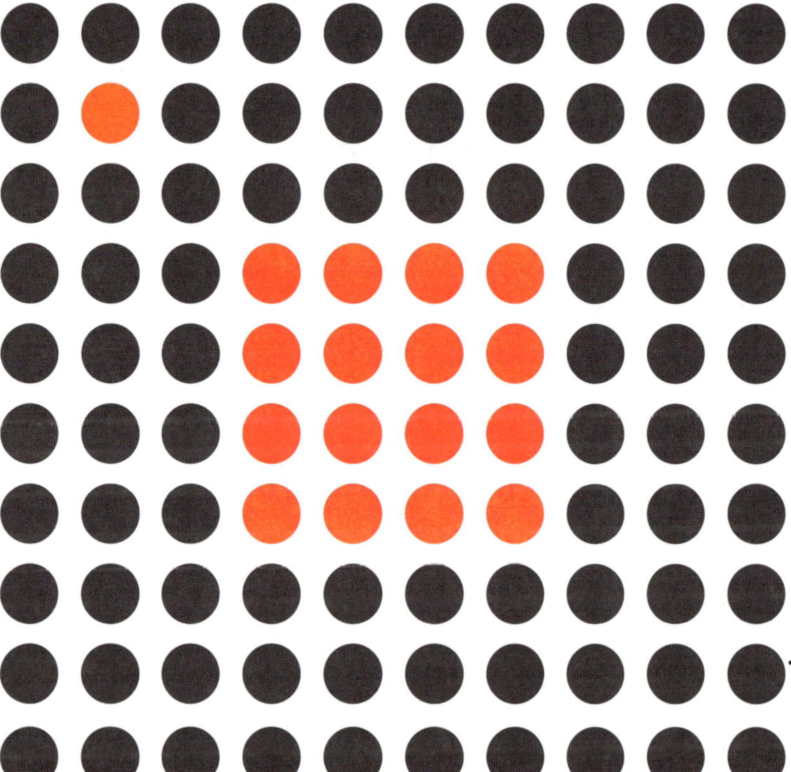

17

ON AND ON AND ON AND ON AND ...

The fraction $\frac{1}{2}$ can be written as a decimal 0.5. Similarly, $\frac{1}{4} = 0.25$. But not all fractions have decimal expansions that stop neatly after a few places. $\frac{1}{17}$ for example produces a repeating decimal

$\frac{1}{17} = 0.0588235294117647\ 0588235294117647\ 0588235294117647\ \ldots$

The 16-digit string 0588235294117647 is called the 'repetend' of $\frac{1}{17}$. For more of these beautiful decimals see chapter 97.

THE AMAZING NUMBER 153

1 + 2 + 3 + 4 + ... + 16 + 17 = 153, which by itself isn't that amazing. But 17 has given us an introduction to 153, one of my favourite 3-digit numbers. For what looks like such an ordinary number, 153 has a load of amazing properties:

$153 = 1^3 + 5^3 + 3^3$ making it 'narcissistic' (see chapter 39)

$153 = 1! + 2! + 3! + 4! + 5!$

The sum of the digits of 153 is a perfect square (1 + 5 + 3 = 9) and the sum of all of the factors of 153, apart from 153 itself, which we call the 'aliquot' parts of 153, is a perfect square ($1 + 3 + 9 + 17 + 51 = 81 = 9^2$).

Not only that, 153 is a Harshad number (see chapter 84) because it is divisible by 1 + 5 + 3 = 9.

Add 153 to its reverse and you get 153 + 351 = 504.

Further, $504^2 = 288 \times 882$; so, 504 is the sum of 153 and its reverse and its square is equal to the product of a number and its reverse. And while we're on the topic of reversing the digits of 153, we started with:

153 = 1 + 2 + 3 + 4 + ... + 17 and 351 = 1 + 2 + 3 + 4 + ... + 25 + 26.

And just to finish us off, $153 = 3 \times 51$.

Nice work, 153.

SOLVABLE SUDOKUS

Number puzzles appeared in newspapers in the late 19th century, when French puzzle setters began experimenting with removing numbers from magic squares, but the phenomenon known as 'Sudoku' exploded in popularity in the 2000s.

The aim is to place the numbers 1 to 9 in every row and column and 3×3 square of a grid – with no repetition of any number in a row, column or 3×3 square.

In 2005, Bertram Felgenhauer and Frazer Jarvis showed that there are $9! \times 72^2 \times 2^7 \times 27{,}704{,}267{,}971 = 6{,}670{,}903{,}752{,}021{,}072{,}936{,}960$ different valid Sudoku grids. But lots of these grids are essentially repeats of each other; 2 grids might be the same under reflection or rotation, or if you swap all the ones and threes in a grid, or if you swap some rows and columns. When you remove these repeats, you get 5,472,730,538 unique solutions.

The Sudoku puzzle to the right has only 17 'givens' and for a long time it was thought that you couldn't create a Sudoku with only 16 givens that had a unique solution.

							1	
					2			3
			4					
						5		
4		1	6					
		7	1					
	5					2		
				8			4	
	3		9	1				

Believe it or not, over 50,000 examples of solvable Sudokus with 17 givens had been found and no such 16 given Sudokus. But that weight of evidence in itself was not a proof.

Congratulations to Gary McGuire, Bastian Tugemann and Gilles Civario who proved in 2012 that there is no solvable 16-clue Sudoku.

ODD FACT

$17^3 = 4913$ and $4 + 9 + 1 + 3 = 17$. So 17 is a number for which the sum of the digits of its cube is the number itself.

QUIZ QUESTION: Apart from the trivial case of 0 and 1, this also happens for a 1-digit number, two numbers in the teens and two numbers in the 20s. Can you find all five numbers for which the um of the digits of its cube is the number itself? ANSWER AT THE BACK OF THE BOOK

BLACKJACK

In a hand of blackjack or 21, the dealer must keep drawing until their hand totals at least 17.

In some casinos, if the dealer has a 'Soft 17', that is, a score of 17 involving an ace, they must draw again. Best you check if that's the rule before you start mouthing off!

While there are many prime numbers that are the sum of 3 successive primes, for example, $5 + 7 + 11 = 23$, $11 + 13 + 17 = 41$; and 5 successive primes, for example, $11 + 13 + 17 + 19 + 23 = 83$, $31 + 37 + 41 + 43 + 47 = 199$; there is only *one* prime number that is the sum of 4 successive primes. You guessed it:

$$17 = 2 + 3 + 5 + 7$$

And 17 is the smallest number that can be written as the sum of a square and a positive cube in two different ways.

QUIZ QUESTION: Find the two different ways to express 17 as the sum of a square and a positive cube. If you're feeling really brave, find the three ways we can express 17 as $a^2 - b^3$. ANSWER AT THE BACK OF THE BOOK

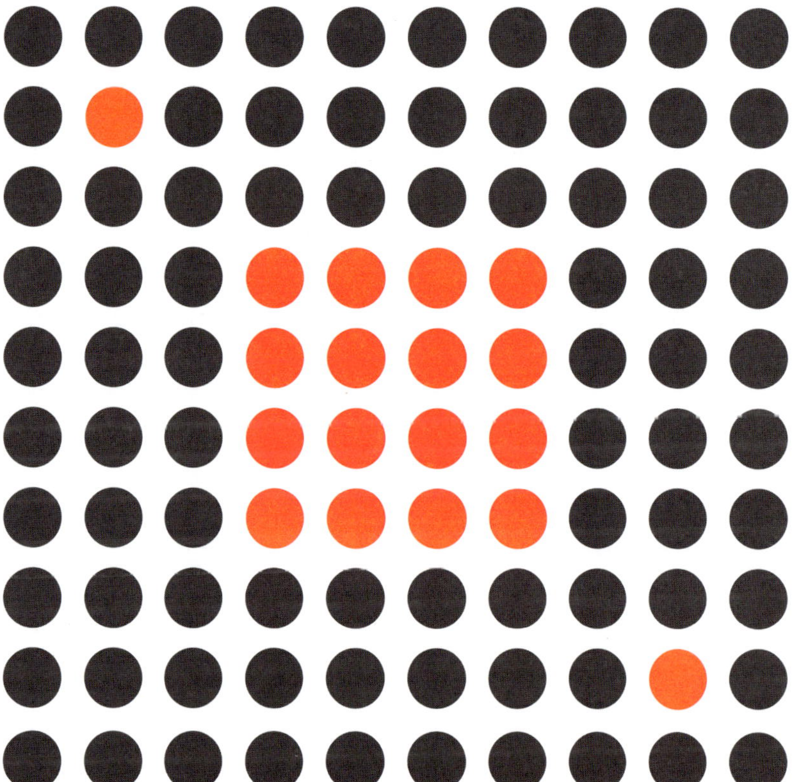

18

Eighteen is the 6th Lucas number. The Lucas numbers start with 1 and 3, then, following the same pattern as the Fibonacci numbers, we add the last 2 numbers to get the next one. So, 1 + 3 = 4, 3 + 4 = 7, giving us the Lucas numbers 1, 3, 4, 7, 11, 18 ... This isn't the only connection between the Lucas and Fibonacci numbers. If you're keen, write $L_1 = 1$, $L_2 = 3$, $L_3 = 4$ ... for the Lucas numbers and $F_1 = 1$, $F_2 = 1$, $F_3 = 2$ ... for the Fibonacci numbers and check out these little gems: $L_n = F_{n-1} + F_{n+1}$, $Ln = Fn + 2 - F_{n-2}$ and $F_{2n} = L_n \times F_n$

THE TRUNCATED TETRAHEDRON

... has 8 faces, 12 vertices and 18 edges. Check that Euler's formula V + F − E = 2 applies to this Archimedean solid.

GOLF COURSES

Golf courses have 18 holes – though many of my friends claim they do their best work at the 19th.

Speaking of golf and 19 – which is 91 backwards – my Uncle Jack shot a hole in one at his beloved Tuggerah Lakes golf course on the central coast of NSW Australia at the age of ... you guessed it ... 91.

Legend.

A RICKETY OL' PROBLEM

Time for you to be tormented by a famous mathematics problem that has given a lot of people headaches over the years including me when I first saw it.

Amy, Ben, Cassius and Delia need to cross a rickety old wooden bridge between two mountain tops late at night. Don't ask why, that's not important. They only have one very weak torch which means only two of them can cross at any time. In addition, when two people cross together, they have to travel at the slowest person's pace so they can both see. And to top things off, in exactly 18 minutes the bridge will collapse (again, see 'don't ask why, that's not important').

Amy can get across the bridge in 1 minute, Ben 2 minutes, Cassius takes 5 minutes and Delia 10.

If Amy and Delia go across, that takes 10 minutes and Amy returns with the torch taking 1 minute.

If Amy and Cassius then cross and Amy returns, it takes 5 minutes plus 1 minute. Amy and Ben then cross together in 2 minutes.

In total, that takes 10 + 1 + 5 + 1 + 2 = 19 minutes and just about anyone you ask comes up with that as the quickest possible crossing time. Like I said I certainly did when I first saw the problem.

But in fact they can get across quicker. Not in 19 minutes ... not in 18 minutes, but under 18 minutes.

QUIZ QUESTION: Find a combination of crossings that gets Amy, Ben, Cassius and Delia safely across the river in less than 18 minutes.

HINT: There's no 'trick' involving three people crossing at once or anything; two people can cross and one return and so on like in the 19-minute example but with the right combinations of people it can be done in less than 18 minutes. Honest!

ANSWER AT THE BACK OF THE BOOK

PEG SOLITAIRE

The game peg solitaire – also called Hi-Q – was the sort of game that was very popular back in the day (when anyone says 'back in the day', basically they mean 'before the internet, mobile phones, Facebook and anything else people over the age of 40 are afraid of').

To play the most common form of peg solitaire, you start with 32 pegs and one empty hole, as in the diagram below.

You jump a peg over any single adjacent peg and into a hole (adjacent means not diagonally, only horizontally or vertically) and remove the peg you've jumped over. If one peg can do several jumps in a row, that is considered a single move.

The idea is to end up with a single peg in the middle square.

In 1912, the English writer Ernest Bergholt found an 18-move solution.

A quick search on the web for 'peg solitaire' or 'Ernest Bergholt' will provide said solution and more information than even I really want about this fascinating game.

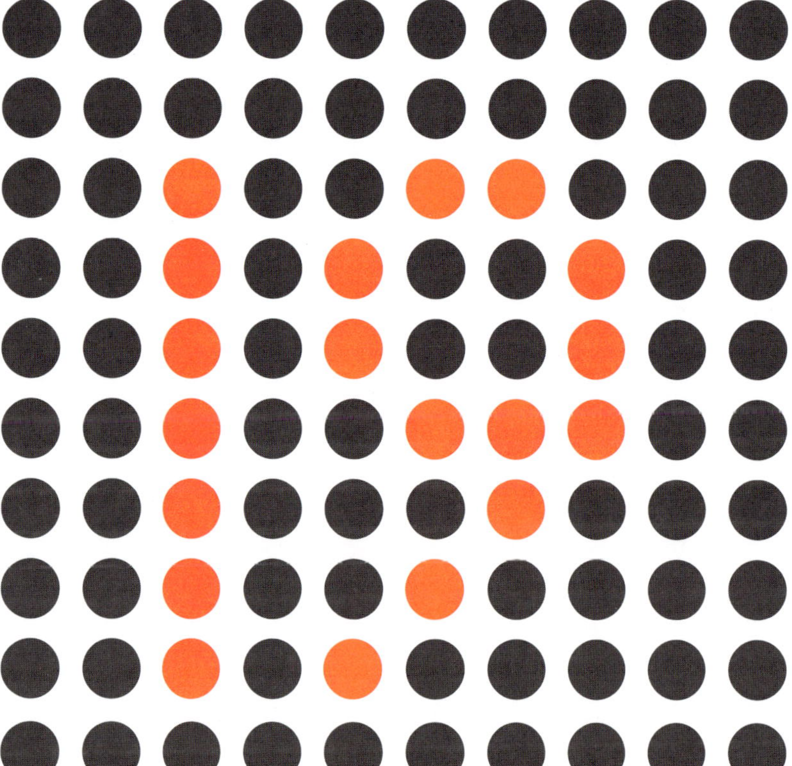

19

GO!

The Chinese board game Go, also known as *igo* (in Japanese), *weiqi* (in Chinese) or *Baduk* (in Korean), is thought to be one of the most strategically complex of all board games, especially given how simple are its rules.

Put simply, you attempt to surround your opponent's pieces, thus capturing the pieces that you have surrounded and claiming that territory. The winner is the player who has seized the most territory by the end of the game.

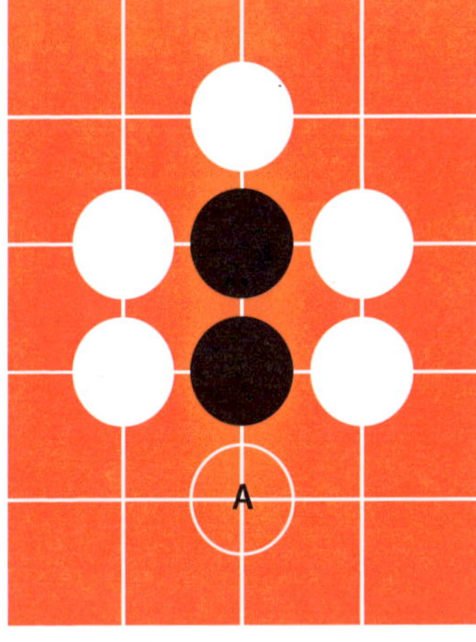

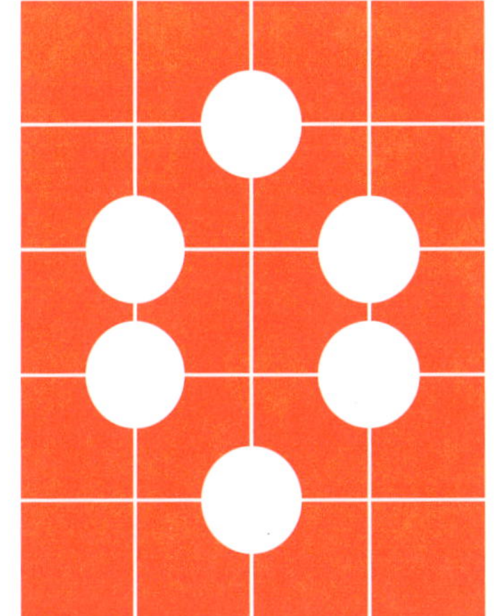

For example, if white puts a piece at position A (above), they capture the two black pieces inside.

The Go board is a grid of 19 × 19 lines and stones are placed on the intersection of the lines. For a game with such simple rules, it requires remarkably deep strategic thinking. While a game of chess can lead to about 10^{120} possible board configurations, Go has in the vicinity of 10^{170}. Even though these are both ridiculously large numbers, it could be argued that Go has 100,000,000,000,000,000,000,000,000,000,000,000,000,000,000,000,000,000 as many possible ways of being played as chess!

Originally, the grid of lines on the board was 17 × 17. It's thought by the experts that a 21 × 21 grid would work but the tactics would become so complicated that barely a handful of players would be able to compete. In the famous book, *The Game of Go*, Arthur Smith writes, 'If more than 21 lines are used ... the combinations are beyond the reach of the human mind'.

BERRY'S PARADOX

Time for a paradox, so take a deep breath.

The number 11 can be described as 'e-lev-en' or as 'six plus five', both of which use 3 syllables. But it's not possible to describe or name 11 in only 2 syllables.

If you think about the numbers 1 to 10, you'll see that 11 is the smallest integer not nameable in fewer than 3 syllables.

It follows that there must be a number out there, bigger than 11, that is the least integer (or whole number) not nameable in fewer than 4 syllables, and one for 5 syllables and so on.

The British philosopher Bertrand Russell once wondered about 'the least integer not nameable in fewer than 19 syllables'.

Why did he pick 19? Because his statement has exactly 18 syllables. So whichever is the least integer that can't be described in fewer than 19 syllables can actually be so described, using Russell's phrase. This number both can and can't be described this way!?

I need a lie down now.

'Bertie' Russell as a young boy, from his Aunt Agatha's photo album. Source: Public Domain

F_{19} IS NOT PRIME

Mathematicians often see a pattern and start to get excited: 'I might be onto something here' they think, dreaming of their name up in lights alongside Einstein and Pythagoras ... before, usually, they crash back to Earth in a sickening thud of reality when their pattern turns out to be not as impressive as they'd hoped.

I can still clearly remember doing my honours thesis on Fermat's theorem when I 'discovered' this:

We all know $3^2 + 4^2 = 5^2$.

Well, it turns out that $3^3 + 4^3 + 5^3 = 6^3$ $(27 + 64 + 125 = 216)$.

'Here we go!' I thought. 'I've stumbled onto some incredible hidden pattern in the powers of integers. All I need to do is prove this more generally and I'll be ...' I didn't even get as far as imagining the word 'famous' before I realised that the next logical step in this amazing pattern I'd discovered ... *wasn't*.

Sure $3^2 + 4^2 = 5^2$ and $3^3 + 4^3 + 5^3 = 6^3$, but $3^4 + 4^4 + 5^4 + 6^4 = 2258$ and $7^4 = 2401$. Not even close.

Turns out I was one of thousands to have fallen into that little trap. We've already met the Fibonacci numbers (see chapters 3, 8, 13). Write the *n*th Fibonacci number as F_n. Look at every prime numbered position on the Fibonaccis – apart from $F_2 = 1$. You'll see that:

$F_3 = 2$, $F_4 = 3$, $F_5 = 5$, $F_6 = 8$, $F_7 = 13$, $F_8 = 21$, $F_9 = 34$, $F_{10} = 55$, $F_{11} = 89$ ...

Here we go then ... is it just me, or apart from F_2, are F_{prime} all prime?

$F_{13} = 233$ (I'll be on the front of *Time* magazine ...), $F_{17} = 1597$ (They'll name a planet after me ...)

QUIZ QUESTION: Find the 19th Fibonacci number and show that it *isn't* prime. ANSWER AT THE BACK OF THE BOOK

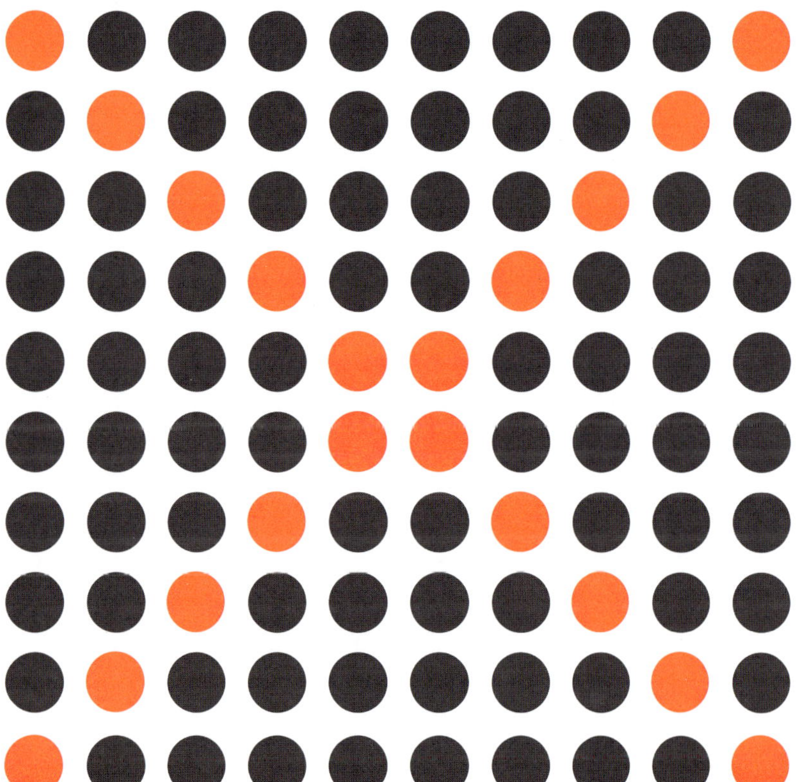

20

20/20

'20/20 vision' is the standard term for normal vision. To have 20/20 vision means that you can see at 20 feet what someone with 'normal' vision sees. That's why it's called 6/6 vision in some places (6 metres).

The best human vision recorded is around 20/8, though 20/10 is exceptional. It is thought the hawk and other birds of prey may have vision as good as 20/2.

Almost all of us at some stage have looked at a Snellen chart but I bet you don't know why. A Snellen chart is the standard eye chart with a large letter followed by smaller letters down the page. It is used to check if you have 20/20 vision.

... just a typical Snellen chart I'd like you to take a long, hard look at ...

T

HI

SBOO

K IS AWES

OME I SHOULD GET

ALL MY FRIENDS TO BUY IT

The Ancient Mayans of Central America used 20 as the basis for their number system and one of their many calendar was made up of 13 months of 20 days each.

GOD'S NUMBER

One of the biggest selling and coolest toys of all time would have to be the Rubik's Cube. Six faces, 6 colours, and over 43 million, million, million possible arrangements, like so many beautiful puzzles and works of art, it is ... blah blah blah ...

But every one of these 43 million, million, million possible arrangements can be solved. So if you wrote down a list of all the possible cube arrangements and beside each arrangement, wrote the number of moves it takes to solve that cube, you'd be able to look at that list of numbers and find out 'the most moves required to solve any possible arrangement of a Rubik's Cube'.

Obviously drawing up the list would be a problem. If you could draw a diagram for a cube and write down the move number in just one second, the list would take 81,000,000,000,000,000,000 years – which could slow down the whole process a bit.

Regardless, this number is called the 'God number' by many mathematicians – the joke being that if God exists, he or she could certainly solve any Rubik's Cube in as few moves as possible.

In 2011, American cuber Tomas Rokicki and his pals proved that any Rubik's Cube can be solved in at most 20 moves. God's number is 20. They did this by reducing the 43,252,003,274,489,856,000 possible arrangements of a cube into 55,882,296 'typical' cases and solving each of them. This took effectively 35 CPU-years of idle computer time donated by the nice people at Google Labs.

A good cause, if you ask me.

ICOSAHEDRON

A regular icosahedron has 20 faces, all of them equilateral triangles (triangles with all 3 sides of equal length). It is one of the 5 Platonic solids all of which were known to the Greek philosopher Plato as far back as 350 BC. We know this because he mentioned them in a handy little page-turner called *Timaeus* that he banged out around this time.

While having an extraordinary grasp of geometry for the time, Plato also thought that the icosahedron was the shape of the smallest occurring particles of water (which he considered one of the fundamental elements or building blocks of everything in the universe). On this score, Plato was proven to be, how shall we say, 'not quite so correct', but hey, that's easy for us to say in our modern world with our swag electron microscopes and the like.

Having 20 faces gives the icosahedron the most faces of any of the Platonic solids.

You might want to check Euler's formula for polyhedra (see chapter 30) on these bad boys. Then again, maybe you don't? Your loss.

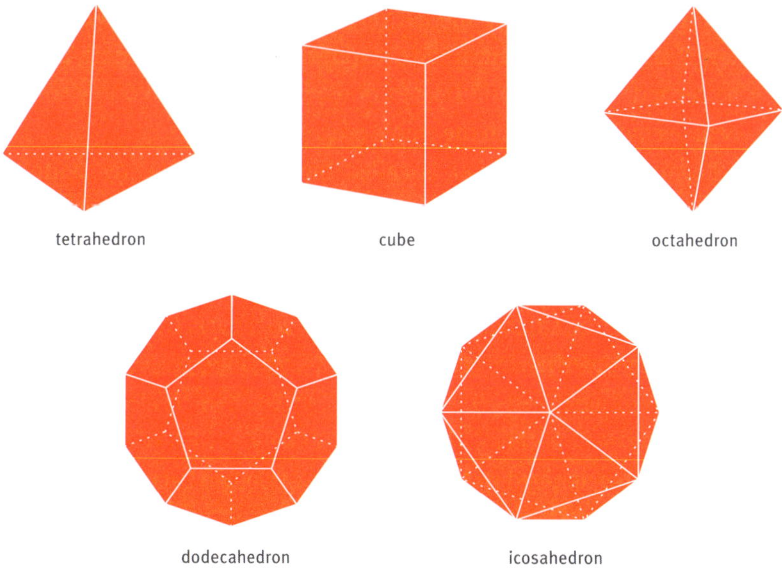

21

21 SQUARES

You can easily divide a square into 4 smaller squares of equal size. But if the smaller squares have to all be *different* sizes, you need 21 squares, and there is only one way you can do it. It was found by Dutch computer scientist and mathematician A. J. W. Duijvestijn in 1978.

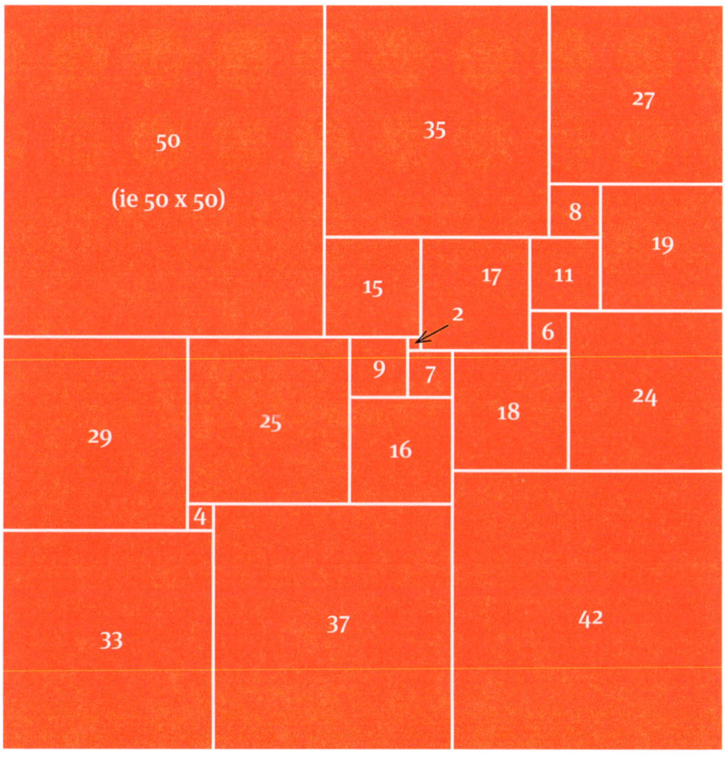

While some crazy lawn bowlers like to change it up a bit, according to the Laws of the Sport of Bowls, in a game of singles, the winner is the first to 21 points.

But in 2001, in an effort to modernise the game, the International Table Tennis Federation changed a set from first to 21 with 5 serves each to first to 11 with 2 serves each.

TRIANGULAR NUMBERS

Twenty-one is the 6th 'triangular' number, and hence the total number of dots on a standard die.

Triangular numbers come from the number of dots in a right-angled triangle diagram like this:

They are formed by adding up the series 1 + 2 + 3 + 4 + 5 ... Some triangular numbers are also square numbers (36, for example). But, apart from the trivial case of T_1 no triangular number can be a cube or a 4th or a 5th power. To find out the *n*th triangular number or T_n, use the formula:

$$T_n = \frac{1}{2} \times n\,(n+1)$$

$$\begin{aligned} \text{So: } T_6 &= \frac{1}{2} \times 6\,(6+1) \\ &= \frac{1}{2} \times 6 \times 7 \\ &= 21 \end{aligned}$$

African
elephants
have 21
pairs of
ribs.

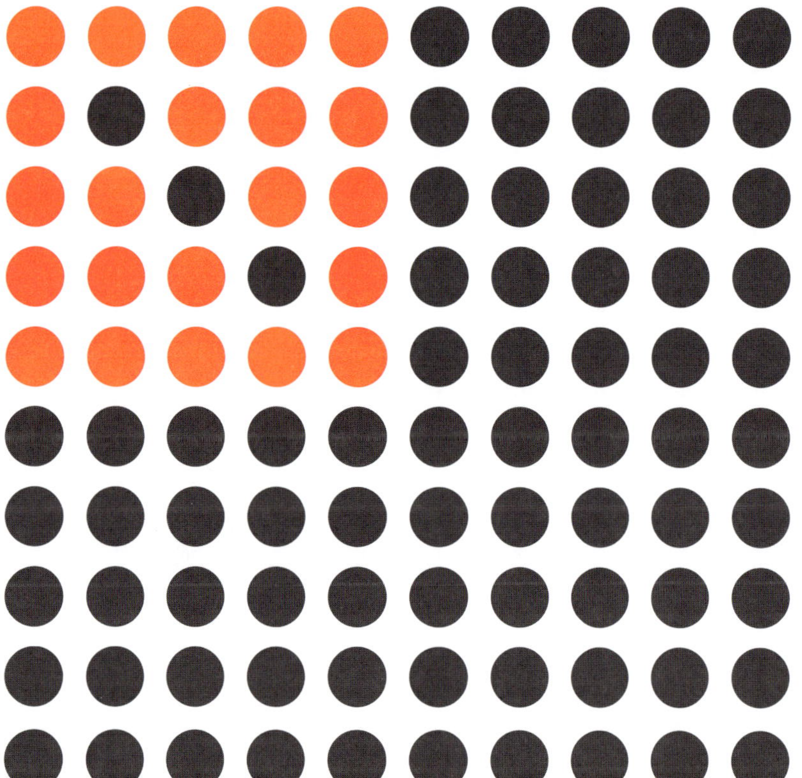

22

There are 22 partitions (see chapter 4) of the number 8 – feel free to find them all.

PIECE OF PI

Probably the most famous number of all is π – pi – the ratio of the circumference of a circle to its diameter. In real words, this means that if a circle is 1 metre across, then it's π metres around the outside.

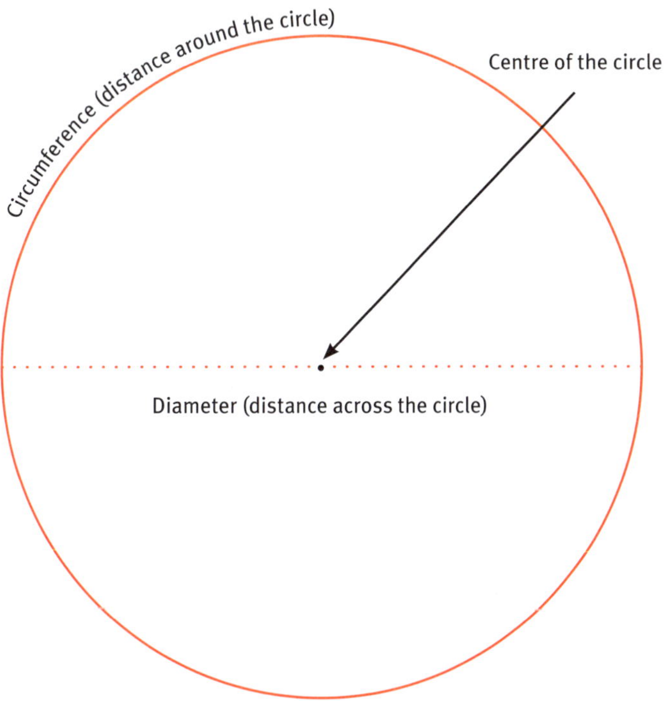

But π is an irrational number, so we can't write it as a fraction or exact decimal. The best we can do is make an approximation of it. As a decimal, π equals roughly 3.14159. And this is where 22 comes in: the most commonly used fraction approximation of $\pi \approx \frac{22}{7}$. So if a circle is 7 centimetres across, it's roughly 22 centimetres around.

MORE TASTY PI RECIPES

$\frac{22}{7}$ is a handy approximation for pi but it is not equal to pi. So it makes sense that there are other fractions out there that are even better approximations to pi – and you can write pi as

$\frac{333}{106}$ or $\frac{355}{113}$ ... and if you really want to impress, try $\frac{52163}{16604}$ or $\frac{103993}{33102}$ which is 99.99999998% accurate.

Fractions aren't the only way we can approximate pi. We can use all manner of expressions that are either gorgeous or scary looking depending on your point of view:

$$\pi \approx \sqrt{7+\sqrt{6+\sqrt{5}}}$$

... gives pi to 3 decimal places and:

$$\pi \approx \frac{99^2}{2206\sqrt{2}}$$

... to 7 places.

And pi can also be written many ways as an infinite series (see chapter 1) in particular:

$$\frac{\pi}{4} = \sum_{k=0}^{\infty} \frac{(-1)^k}{2k+1} = 1 - \frac{1}{3} + \frac{1}{5} - \frac{1}{7} \dots$$

and:

$$\frac{\pi^2}{6} = \sum_{k=1}^{\infty} \frac{1}{k^2} = \frac{1}{1^2} + \frac{1}{2^2} + \frac{1}{3^2} + \dots = \frac{1}{1} + \frac{1}{4} + \frac{1}{9} + \dots$$

Where the freaky looking sigma sign (Σ) is maths notation for the sum of a series of terms. Wow, that's a lot of pi. Sorry – couldn't resist.

THE WHOLE 22 YARDS

A cricket pitch is 22 yards long – but must have felt a lot shorter when Australian fast bowler Fred 'The Demon' Spofforth had a head of steam up. In 1882 at the Oval in England, Spofforth pretty much singlehandedly routed the home team from a seemingly unloseable position. England lost their last 6 wickets for 12 runs and Spofforth took 14 wickets for the match. To commemorate the death of English cricket, the bails were burnt and interred in an urn and one of sport's greatest rivalries, The Ashes, was born.

PENTAGONAL NUMBERS

Remember triangular numbers (see chapter 15)? Well, after triangular numbers and squares come 'pentagonal' numbers, from the 5-sided figure the 'pentagon'. The idea is the same: we keep expanding by putting another pentagon on the outside. The pentagonal numbers are the total number of points in each of the diagrams.

The diagram below shows us that the first 4 pentagonal numbers are 1, 5, 12, and 22.

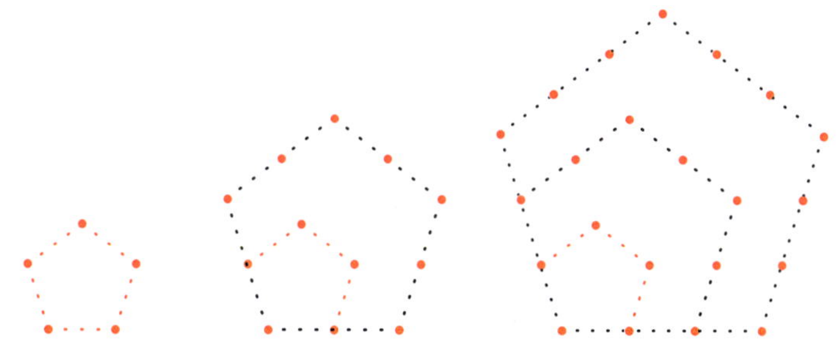

23

THE BIRTHDAY PROBLEM

If there are 23 or more people in a room, the odds of 2 people sharing a birthday are better than 50:50. For fewer than 23 people, it's more likely that nobody will share a birthday. As the 'tipping point' for this problem, 23 is a surprisingly small number. Most people to whom I explain the question imagine you'd need a group of 80 or 100 people, but no, it's just 23.

In fact, once you get above 40 people, a shared birthday is very likely (above 90%), and in a group of 50 people there is a staggering 97% chance.

I'll show you that 23 is the break-even point for the shared-birthday party trick. In fact, as is often easier in probability questions, we'll work out the odds of the exact opposite happening.

Take one person. They have a birthday on some day – say, 1 January. Along comes a second person. The odds that the second person's birthday does not clash with the first person's is:

Jan 1 Jan 29

$\frac{364}{365}$

... because there are 365 days in the year and 364 of those days are not 1 January.

Let's say that this second person is born on 29 January (which is actually my birthday – no need to send a gift).

These two people hang out chatting about what it's like not to share a birthday when along comes a third person.

The odds that their birthday falls on a separate day again is $\frac{363}{365}$, because there are 363 days in a year that aren't Jan 1 or Jan 29.

So, in a group of 3 people, the chance of the second person not clashing birthdays with the first, and the third person not clashing with either of the others, is:

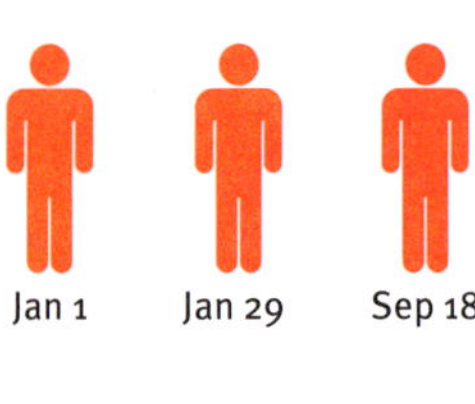

$\frac{364}{365} \times \frac{363}{365}$

... for a group of 4, we follow the same logic and the odds are:

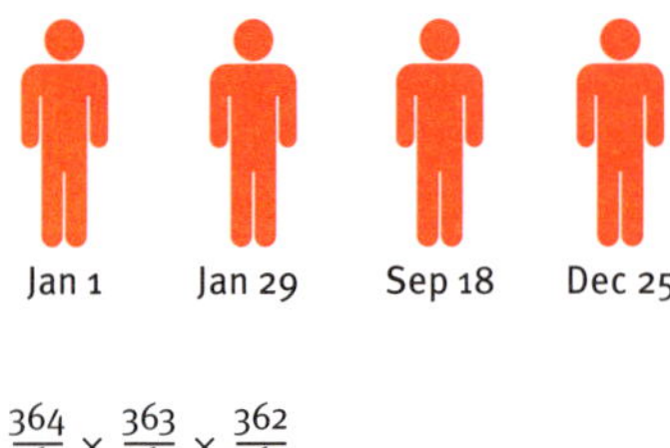

$\frac{364}{365} \times \frac{363}{365} \times \frac{362}{365}$

... that there is no common birthday. Can you see the pattern?

For larger groups, we just keep multiplying terms on the end. So if you have a group of 23 people, the odds of no shared birthdays are:

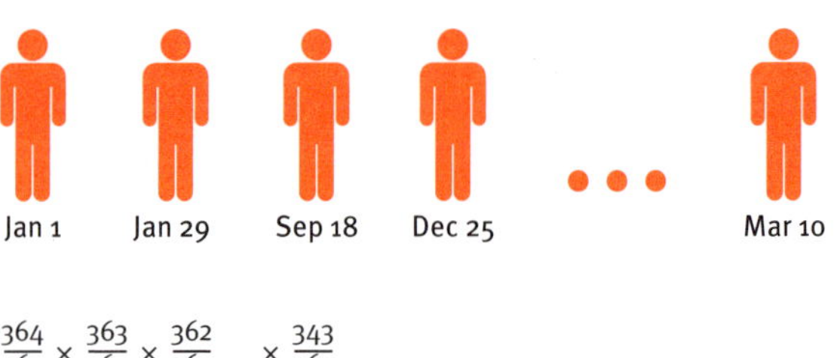

$\frac{364}{365} \times \frac{363}{365} \times \frac{362}{365} \dots \times \frac{343}{365}$

... which is about 0.49. So there is a 51% chance of a shared birthday!

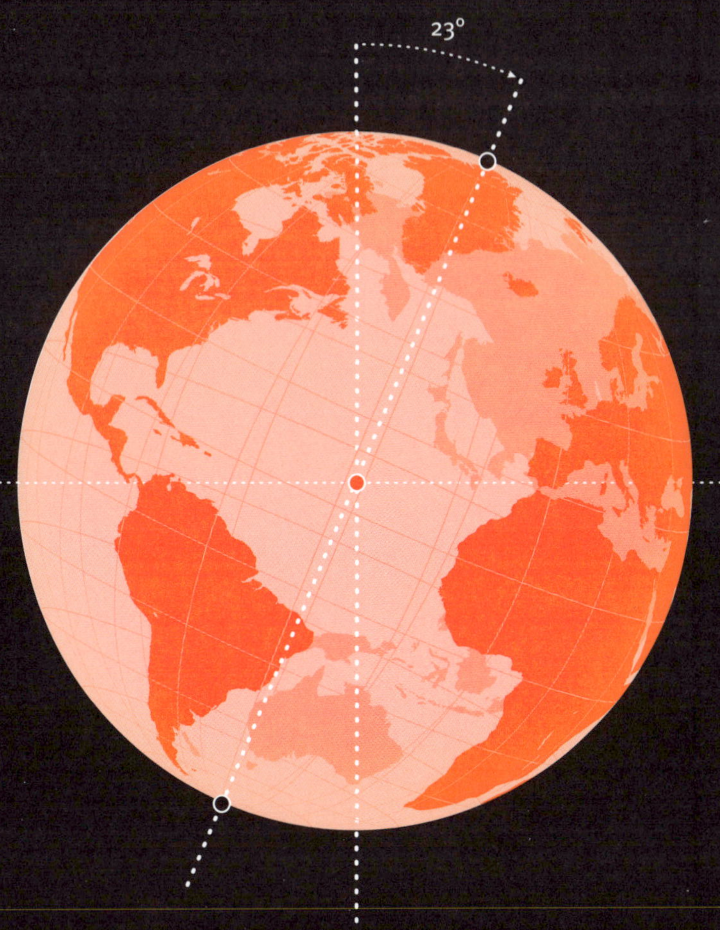

As the Earth travels around the Sun, it also rotates about its axis (an imaginary line from the North to South Pole through the Earth's centre). But the axis of the Earth is actually tilted 23°, which is why the Earth has seasons. If there was no tilt, the hottest and coldest times of the year would be the same for both hemispheres.

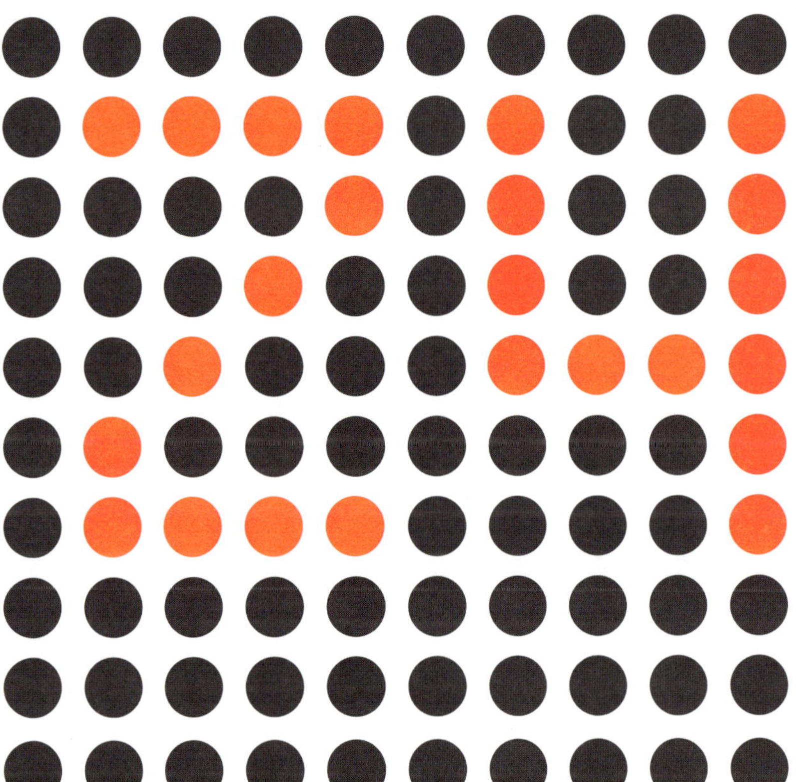

24

Because $24 = 4 \times 3 \times 2 \times 1$, we call it 4 'factorial' or 4! Factorials occur all the time in mathematics, especially in probability and counting problems. Say I had 3 different books and I put them in a pile. There are 3 possible books that could go on the bottom, then 2 choices for the middle, and 1 book left to go on top. So there are $3 \times 2 \times 1 = 3! = 6$ ways of making the pile. By the same thinking, 4 books can be stacked $4 \times 3 \times 2 \times 1 = 24$ different ways.

24 SQUARES

The sum of the first 24 squares is itself a square. Huh?

That means $1^2 + 2^2 + 3^2 + ... + 24^2 = 1 + 4 + 9 + 16 + ... + 576 = 4900 = 70^2$.

Apart from the trivial cases of 0 and 1, 24 is the only number for which this happens.

A big shout out to the awesomely talented Aussie mathematician wunderkind Max Menzies who reminded me of this.

24 POINTS

There are 24 points on the board in a game of backgammon. Each player begins with 15 pieces which they must try and remove from the board before their opponent does the same.

Backgammon is a game played with great passion by its followers around the world – cool backgammon terms include 'lover's leap' (an opening roll of 6 and 5); 'anti-joker' (a very bad roll in a given situation), and a 'Barabino' (a roll of 5–4 from the bar used to make an anchor on your opponent's five-point). Huh?

A SMALL RHOMBICUBOCTAHEDRON

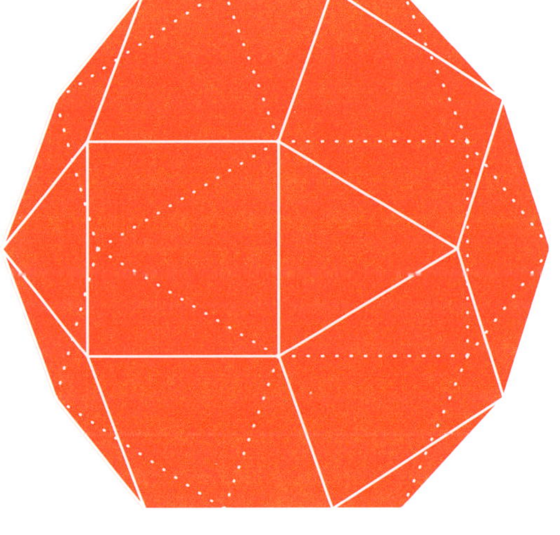

... has 26 faces, 24 vertices and 48 edges! Check that Euler's formula $V + F - E = 2$ applies to this Archimedean solid.

If you're interested, a quick search online will turn up free diagrams to make all the Archimedean solids. A great activity if you're feeling crafty. Below is as an example to get a small rhombicuboctahedron rolling. Makes a great table decoration for a particularly nerdy dinner party.

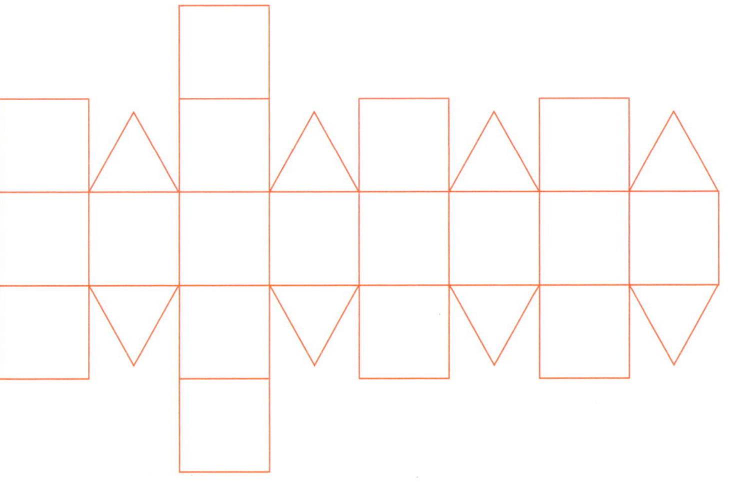

THE FOURTH DIMENSION

We live in a world with 3 dimensions (3D) of space. As such, we can picture things that are 3 dimensional, or 1 or 2 dimensional fairly easily. For example, we can see that a box has width, depth and height. And we can imagine that a piece of paper which has no depth at all would therefore be 2 dimensional. We can see that a thin straight piece of wire goes in one direction but doesn't have very much height or depth and we can therefore picture in our minds a line which is like an 'ultimately thin' piece of straight wire.

But when it comes to picturing objects that exist in 4 dimensions (4D) of space, we understandably struggle.

In the same way that four straight lines can be used to make a 2D square, and six 2D squares can form a 3D cube, in 4D space we can use 3D cubes to assemble the next shape in this series, which we call a 'tesseract' or '4D hypercube'.

While it doesn't help you picture it well enough to be able to spot one at the shops, this diagram helps us count the number of edges, corners and faces of a tesseract.

We picture a 3D cube coming off each of the 6 faces of a 3D cube to get a shape with 24 faces, 32 edges and 16 corners or vertices.

Note that Euler's formula $V + F - E = 2$ from chapter 3 does *not* apply to this 4D shape.

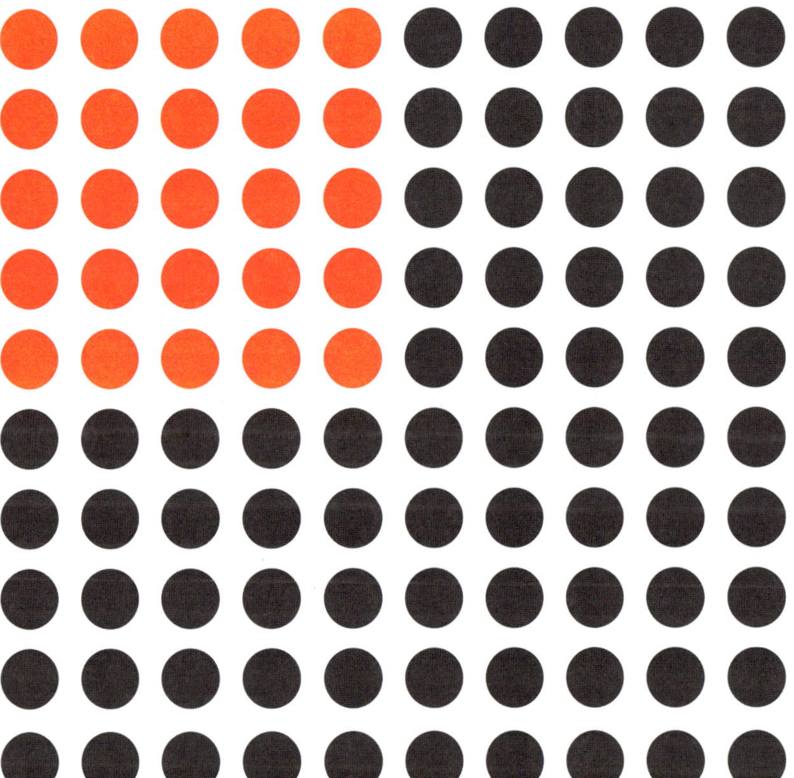

25

LUCKY NUMBERS

Twenty-five is a 'lucky' number. No, I'm not talking some bizarre numerology-aromatherapy gooshta here. To find the lucky numbers, write out a long list of numbers, say 1 to 100:

1 2 3 4 5 6 7 8 9 10 11 12 13 14 15 16 17 18 19 20 21 22 23 24 25 26 27 28 29 30 31 32 33 34 35 36 37 38 39 40 ...

Look at number 2, the second number on the list, and remove it and every second number after it. The list now reads:

1 3 5 7 9 11 13 15 17 19 21 23 25 27 29 31 33 35 37 39 41 43 45 47 49 51 53 55 57 59 61 ...

The next number on the list is 3, so now remove every third number. This would be the 5, the 11, the 17 and so on. The list is now:

1 3 7 9 13 15 19 21 25 27 31 33 37 39 43 45 49 51 55 57 61

Remove every seventh number still on the list and you get:

1 3 7 9 13 15 21 25 27 31 33 37 43 45 49 51 55 57 ...

Remove every ninth number on the remaining list. Continue this pattern, or 'sieve', as mathematicians call it, forever and you're left with the lucky numbers.

QUIZ QUESTION: Find all the lucky numbers up to 100. ANSWER AT THE BACK OF THE BOOK

THE 5 QUEENS PROBLEM

In chapter 12 we met the powerful chess piece, the queen, which can attack squares along any row, column or diagonal on which she sits. We also saw how 8 queens can be arranged on an 8 × 8 board, while being unable to attack each other.

If we were playing on a 5 × 5 chessboard, we could arrange 5 queens 'safely' this way:

QUIZ QUESTION: There is only one other way to safely arrange 5 queens on a 5 x 5 board that isn't this way just spun around or flipped over. Can you find it? ANSWER AT THE BACK OF THE BOOK

Twenty-five is Ireland's national card game and is related to 'ombre', the classic Spanish card game. The Canadian game 'forty-fives' was spawned by 'twenty-five', which itself was called 'Maw' as far back as James I and has also been known as 'spoil five'. I've read it described as a game wherein 'there is a fair amount of luck, but some scope for skill'. Sounds like my type of game.

When we square 25 we get $25 \times 25 = 625$ and because its square ends with the number itself we say that 25 is 'automorphic'. Like that? Cute hey? Now if you work out 24^3 you'll see why we call 24 'trimorphic'.

PERFECT SQUARE

Twenty-five is also a perfect square: $5^2 = 25$. Now, the product of any 4 consecutive numbers + 1 will always equal a square. So:

$$1 \times 2 \times 3 \times 4 + 1 = 25 = 5^2$$

or

$$11 \times 12 \times 13 \times 14 + 1 = 24{,}025 = 155^2$$

For those of you who can multiply algebra, the proof isn't too hard to follow:

$$n(n + 1)(n + 2)(n + 3) + 1$$

which is just 4 consecutive numbers plus 1, will open up into:

$$n^4 + 6n^3 + 11n^2 + 6n + 1$$

... which is just $(n^2 + 3n + 1)^2$.

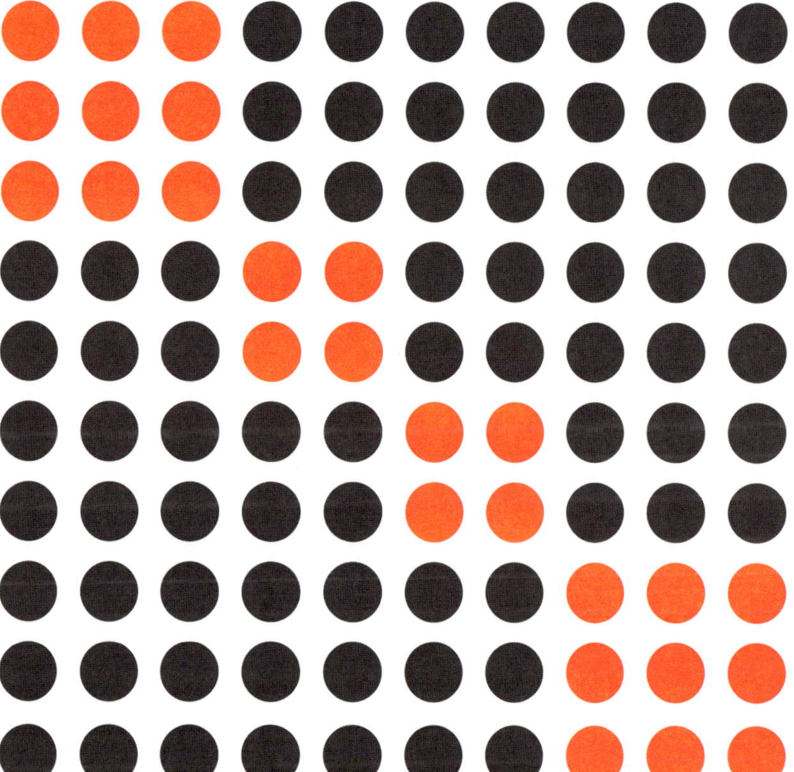

26

CANTONS

Switzerland is divided into 26 'cantons' – including such beautiful names as Obwalden, Valais and Schaffhausen.

MARATHON TIP

Once you've run 26 miles of your first marathon you only have in front of you 385 of the toughest yet most beautiful yards you'll ever run.

$5^2 = 25$ & $3^3 = 27$

Thanks to Euler (who else?), and a few others, we know that 26 is the only number so sandwiched between a square and a cube.

BOSONIC STRING THEORY

Most people think of our world as being 3 dimensional (3D).

In 1905, Einstein convinced us to think of time as the 4th dimension of our world and the Universe. He showed that time and the 3 dimensions of space were intimately connected.

But in the 1960s we saw the emergence of 'string theory' that suggests there are *more* than 3 dimensions of space but that these extra dimensions are so small we can't see them in our everyday lives.

To help you understand this, people sometimes use an analogy of a garden hose that from a long way away looks just like a 1-dimensional line but up close it's possible to move along the line but also actually be moving in other dimensions, for example, a spiral around its edge. Similarly, other dimensions of space might be very tightly 'twisted together' and hard to see.

The original version of string theory was called 'bosonic string theory' and suggested the Universe is actually composed of 26 dimensions of spacetime.

THE LIT ROOM PUZZLE

Imagine that the walls of the room in the diagram below were covered in mirrors. If I stood in the corner of this room and lit a match, the light would bounce around the walls and hit the far corner. In fact, no matter where you stand in the room, the light will eventually bounce around to your eyes.

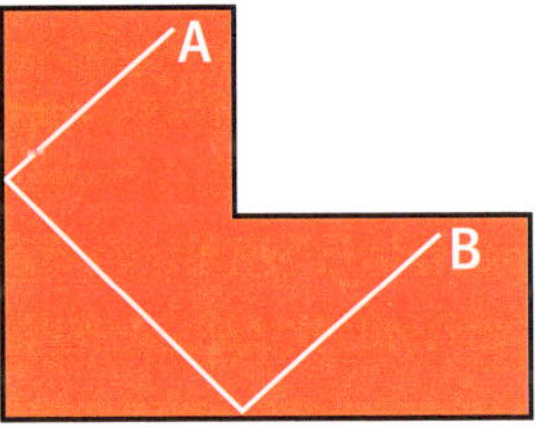

In the 1950s, the talented German-American mathematician Ernst Strauss asked whether it was possible to design a room with so many corners, walls and weird spots that there's a place where you can light a match and another place in the room that can't be lit. Turns out there *is* such a room. In 1995, Canadian mathematician George Tokarsky produced this 26-sided beast:

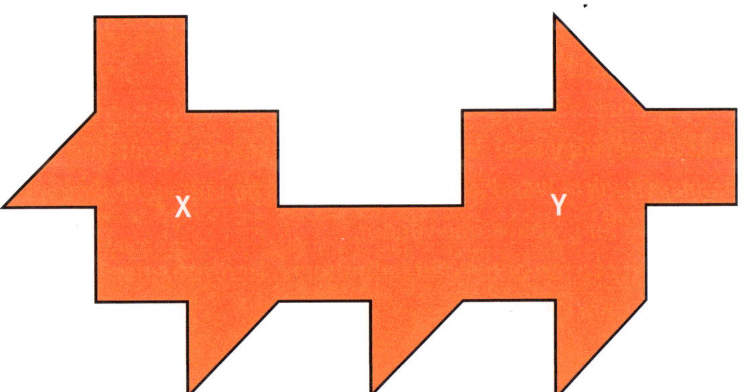

If you light a match at X, a person at Y will remain in darkness.

Since then, a 24-sided room has been found, but no-one can take away from George Tokarsky that he found the first 'unilluminable straight-walled room'.

The modern Roman (or English) alphabet has 26 characters. There used to be only 25, until the letter 'J' came along in the 14th century. Since 1917, the Norwegian (and since 1950, Danish) alphabets have had 29 letters – the 26 of English plus 3 extra characters. The Russian Cyrillic alphabet has 33 characters, while Arabic has 28.

PALINDROMES

Twenty-six is the smallest non-palindromic number to have a palindromic square (676).

On the subject of palindromes (things that read the same from front to back and back to front) Danica McKellar, who played Winnie Cooper in *The Wonder Years* and is now a maths education guru and palindrome lover, gives us this: 'What did the mathematician say when she was offered a piece of cake?'

'I prefer pi.'

Impressive though that might be, American comedian Demetri Martin once wrote a 224-word-long palindromic poem as a project for a fractal geometry class. But even that doesn't come close to David Stephen's 1980 palindromic novel, *Satire: Veritas*, which topped out at 58,795 letters. Dunno whether it's any good, but I'm told you can read it from back to front in one sitting.

And, what, you might be wondering, is the longest palindromic word in everyday use? Well, I'm glad you asked. According to the *Guinness Book of World Records*, that would be *saippuakivikauppias*. It's Finnish for ... soapstone vendor.

I'm a bit of a fan of the odd palindrome. Here are some more:

- Roy, am I mayor?
- A man, a plan, a canal – Panama!
- Satan oscillate my metallic sonatas.
- Straw? No, too stupid a fad. I put soot on warts.
- No misses ordered roses, Simon.
- Dennis and Edna sinned.
- Pusillanimity obsesses boy Tim in *All Is Up*.

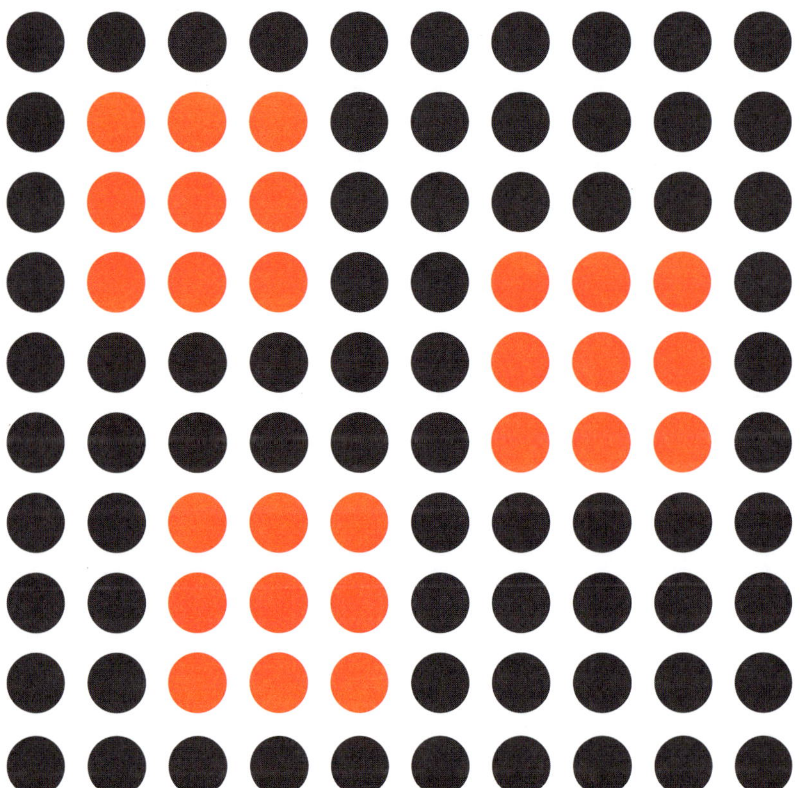

27

Someone emailed me this and I can't find where it came from originally but it's pretty cool:

$$\frac{27^4 + 2^4 + 4^4 + 21^4}{28^4 + 1^4 + 9^4 + 18^4} = \frac{27^2 + 2^2 + 4^2 + 21^2}{28^2 + 1^2 + 9^2 + 18^2} = \frac{27 \times 2 \times 4 \times 21}{28 \times 1 \times 9 \times 18}$$

Convince yourself that this property holds for the numbers (2, 45 , 48 ,91), (7, 9, 78 ,80) and for (3, 35, 54, 92), (4, 23, 63, 90).

CAYLEY'S SEXTIC

How cool is this diagram? It's called 'Cayley's sextic'. Cayley was a brilliant British mathematician in the 1800s. Sextic ... is nowhere near as interesting as it sounds – it just means 'of the 6th power'.

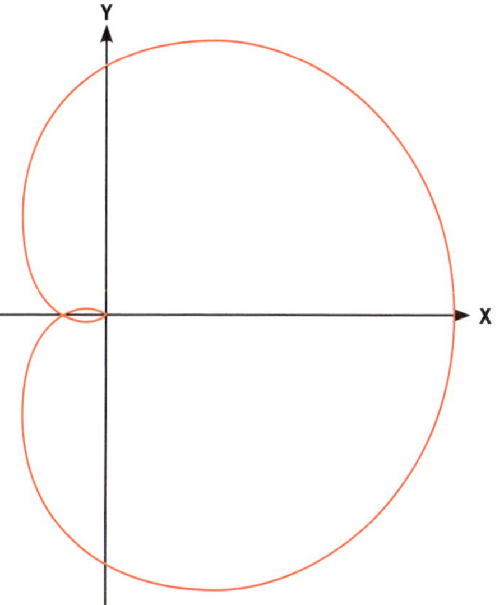

The equation of Cayley's sextic is $4(x^2 + y^2 - ax)^3 = 27a^2(x^2 + y^2)^2$

If you expand all of the brackets, you get a lead term $4x^6$ ... hence sextic.

I've already mentioned what a major gun the Swiss mathematician Leonhard Euler was (surely the name sounds familiar by now!), but, hey, sometimes even he stuffed up.

Consider this:

Right-angled triangles in high school give us the equations $3^2 + 4^2 = 5^2$ and $7^2 + 24^2 = 25^2$ and so on. It's also true that $3^3 + 4^3 + 5^3 = 6^3$.

But we can't get 2 cubes to add up to a cube. Or, more formally, there are no whole numbers *a*, *b* and *c* that solve $a^3 + b^3 = c^3$ (see chapter 67 for a longer explanation).

Euler saw all this going down and suggested that $a^4 + b^4 + c^4 = d^4$ and $a^5 + b^5 + c^5 + d^5 = e^5$ and so on would also have no solutions, or, to use mathspeak, 'You can't write an *n*th power as the sum of less than *n* *n*th powers.' Pretty groovy-sounding, hey!

But in 1966, two mathematicians called Leon Lander and Thomas Parkin showed that $27^5 + 84^5 + 110^5 + 133^5 = 144^5$ ... thus torpedoing Leonhard's really cool conjecture. That's the way it goes, big guy. Even the best of us ...

Leonhard Euler, painted by Jakob Emanuel Handmann in 1756. Source: Public Domain

HAILSTONE PROBLEM

Sometimes in mathematics there is great complexity hidden behind a simple face. Let's play an easy game.

Think of a number, if it is even divide it by 2, if it is odd triple it and add 1. When you get your answer, apply the rule again – halve it if it is even, otherwise triple it and add 1.

Or if you like fancy-pants mathematics talk, apply the function:

$$f(n) = \begin{cases} \frac{n}{2} & \text{if } n \text{ even} \\ 3n + 1 & \text{if } n \text{ odd} \end{cases}$$

For example, 6 – 3 – 10 – 5 – 16 ($16 = 3 \times 5 + 1$) – 8 – 4 – 2 – 1

Or 20 – 10 – 5 – 16 – 8 – 4 – 2 – 1. Let's get a little bit crazy here: 453 – 1360 – 680 – 340 – 170 – 85 – 256 – 128 – 64 – 32 – 16 – 8 – 4 – 2 – 1.

Are you starting to see the pattern?

We think, but it hasn't been proven, that every single number will eventually reduce to 1 – that no matter where you start you will eventually come back to 1 and not rocket off to infinity.

This is called the 'Hailstone problem', the 'Collatz conjecture' or the 'Syracuse algorithm' and was first stated by the German mathematician Lothar Collatz back in 1937.

We have checked this for every number up to 5,764,000,000,000,000,000, but haven't yet proven it for all numbers. I think most people would be flabbergasted if we found a number that didn't come back to 1 but until we've proven it for all numbers mathematicians refuse to say that it's true.

One really cute thing about the Hailstone problem is that two numbers really close to each other can take completely different paths.

QUIZ QUESTION: Try the Hailstone rule starting at 26 then at 28. Show that 26 takes 10 moves to reduce to 1 and that 28 takes 18 moves. Then start at 27. Can you successfully take all 111 steps to get to 1? Feel free to use a calculator or even write a computer program for this one. ANSWER AT THE BACK OF THE BOOK

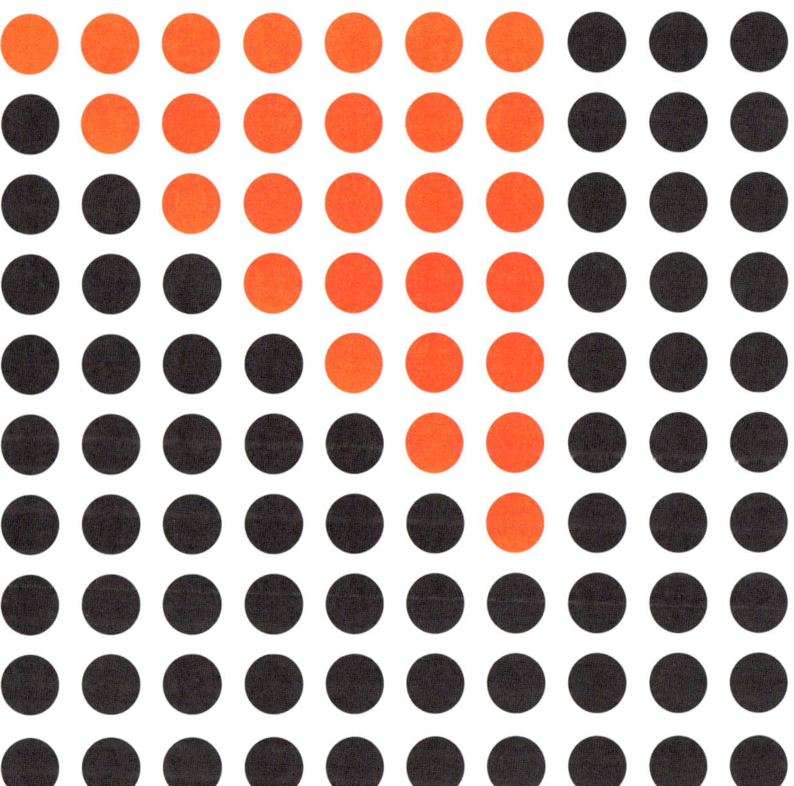

28

ANOTHER PERFECT NUMBER

Twenty-eight is the second 'perfect' number after 6 (if you've forgotten what a perfect number is, go back and take a look at chapter 6) and the seventh 'triangular' number (see chapter 21 if you've forgotten that as well). Here are some more cool things about perfect numbers:

1) All perfect numbers are triangular. For example, $6 = 1 + 2 + 3$

$28 = 1 + 2 + 3 + 4 + 5 + 6 + 7$

The next perfect number $496 = 1 + 2 + 3 + \ldots + 29 + 30 + 31$ etc.

2) We'll meet 'Mersenne' prime numbers in more detail in chapter 31, but for now, these primes and perfect numbers are connected. Any time we have a Mersenne prime – a prime number of the form $2^p - 1$ where p is also prime – we can create a perfect number:

$2^{p-1} \times (2^p - 1)$

So when $p = 2$; $2^2 - 1 = 3$ is a Mersenne prime and

$2^1 \times (2^2 - 1) = 2 \times 3 = 6$ a perfect number.

When $p = 3$; $2^3 - 1 = 7$ is a Mersenne prime and

$2^2 \times (2^3 - 1) = 4 \times 7 = 28$ the second perfect number.

The case $p = 5$ gives us the Mersenne prime 31 and the perfect number 496 and when $p = 7$ we get 127 and 8128 for our Mersenne prime and perfect numbers respectively. You might want to check on a piece of paper that 496 and 8128 are perfect numbers.

In fact, it turns out that every even perfect number has to be related to a Mersenne prime in the way we just showed. You'll never guess who proved that ... okay, yes, it was Leonhard Euler!

3) All the perfect numbers we've ever found have been even numbers. We don't know if there are any odd perfect numbers but if there are they are bigger than the number 1 followed by 1500 zeros. Personally I'd be surprised if we found one – but I've been wrong many times before.

4) A final beautiful thing about perfect numbers is that apart from 6, every even perfect number is the sum of a series of odd cubes:

$28 = 1^3 + 3^3$

$496 = 1^3 + 3^3 + 5^3 + 7^3$

QUIZ QUESTION: Write 8128 as the sum of consecutive odd cubes.

ANSWER AT THE BACK OF THE BOOK

THE FIRST 8 POLYOMINOES

If we ignore the trivial 1 × 1 square, the first 8 polyominoes have a total area of 28 unit squares.

QUIZ QUESTION: Can they all be squeezed into this 4 x 7 rectangle?

ANSWER AT THE BACK OF THE BOOK

DEM BONES

There are 28 bones in the human skull (8 cranial, 14 facial, and 6 ear bones). In total, you should have 206 bones, give or take, made up of the following:

- 28 bones in your skull
- 1 horseshoe-shaped hyoid bone in your neck
- 26 vertebrae (7 cervical or neck; 12 thoracic; 5 lumbar or loins; the sacrum, which is 5 fused vertebrae; and the coccyx, which is 4 fused vertebrae)
- 26 bones in your foot
- 24 ribs plus the sternum or breastbone; the shoulder girdle (2 clavicles or 'collar bones', the most frequently fractured bones in the body, and 2 scapulae)
- 1 pelvic girdle (2 fused bones)
- 3 bones in each arm and 4 in each leg [illegible] fact that each 'arm and hand' has the same nu[illegible] as each leg and foot – 30 bones)
- a few partial bones, ranging from 8 to 18 in number, wh[illegible] are related to joints
- 27 bones in your hand (by comparison, Spider Monkeys [illegible] 45 bones in each hand)

There are individual variations: for example, some people are bo[illegible] with an extra rib or lumbar vertebra and not everyone has Inca (sutural) bones in their skull.

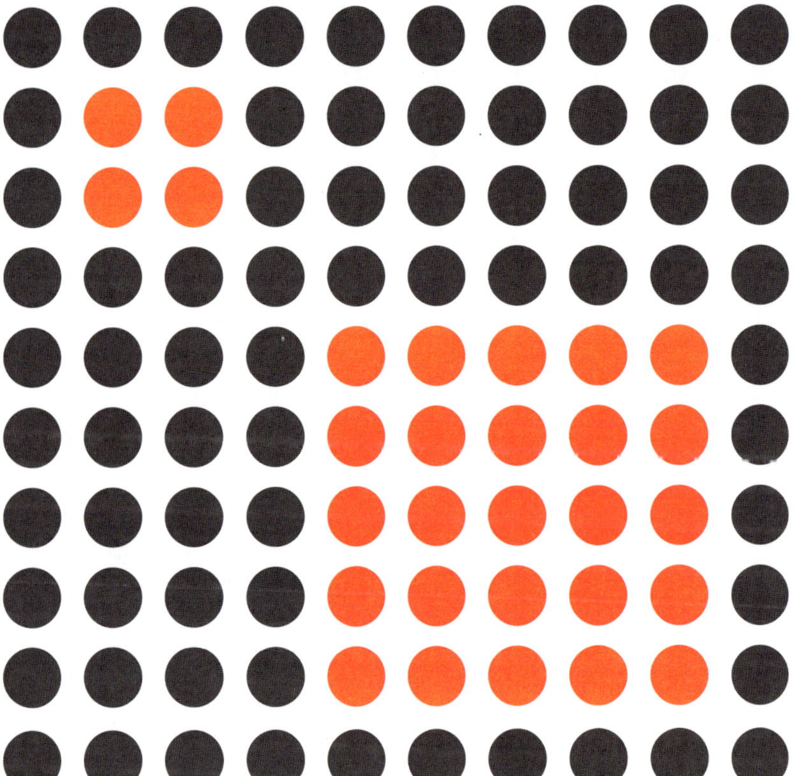

29

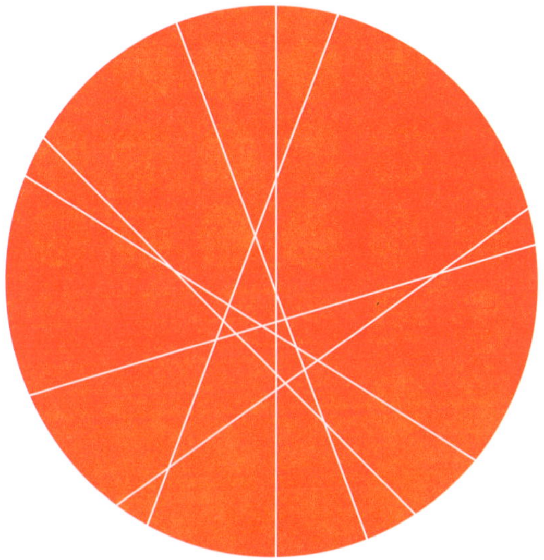

THE LAZY CATERER'S SEQUENCE

Seven cuts through a pizza can create up to 29 pieces. Try it yourself with a pencil and a big circle drawn with a pen. But next time you've got 29 friends coming over, don't think, 'Hey, I only need 1 pizza', because it's hard to work out and some of the pieces will be pretty small. In general, to find out the maximum number of pieces you can get from a certain number of cuts, think of the number of cuts as *n* and put *n* into the formula:

$$\text{pieces of pizza} = \frac{n \times (n + 1)}{2} + 1$$

So with 7 cuts, you could make:

$$\frac{7 \times (7 + 1)}{2} + 1 = \frac{7 \times 8}{2} + 1 = 7 \times 4 + 1 = 29$$

Please note that if you did manage to do this, several of your friends would be *very* dirty about the size and shape of their piece.

PELL NUMBERS

We met 'squares' back in chapter 4 (you'll remember $5^2 = 5 \times 5 = 25$). And as far back as chapter 2, I explained that the reverse of squaring is called 'taking the square root'. So because $5^2 = 25$ we know that 5 is the square root of 25, which we write $\sqrt{25}$.

But the square root doesn't always come out as a nice round number like 5 did in this case. In fact, most of the time it won't.

There is no whole number or fraction that is the square root of 2. But we can work out a list of fractions that get closer and closer to $\sqrt{2}$, and these are called the 'rational approximations of $\sqrt{2}$'.

When we square $\frac{1}{1}$ we get 1, this is less than 2. When we square $\frac{2}{1}$ we get 4, which is further from 2 than 1 was, so 1 is the best approximation we can get for $\sqrt{2}$ for any whole number.

Let's allow ourselves to look at halves as well.

When we square $\frac{3}{2}$ we get $\frac{9}{4} = 2.25$ which is greater than 2 and closer than 1 so $\frac{3}{2}$ is a better approximation of $\sqrt{2}$ than $\frac{1}{1}$ was.

Now nothing with a denominator of 3 or 4 can get closer, but $(\frac{7}{5})^2 = \frac{49}{25} = 1.96$ which is less than 2 and closer again than we got with the previous 'guess' of $\frac{3}{2}$.

The list of best approximations of $\sqrt{2}$ is

$\frac{1}{1}, \frac{3}{2}, \frac{7}{5}, \frac{17}{12}, \frac{41}{29}, \frac{99}{70}, \ldots$

and the denominators of the terms on this list are called the Pell numbers.

So 29 is the fifth Pell number.

29 YEARS

The solar system's second largest planet, Saturn, takes 29 years to orbit the Sun. So you'd really not want to forget someone's birthday ...

29 LETTERS

The Danish and Norwegian alphabets have 29 letters – the 26 we use in English and three others: æ (the uppercase or majuscule of which is Æ), ø (Ø) and å (Å).

$29 = 2 \times 9 + 2 + 9$

and you should be able to convince yourself that any number $a9 = a \times 9 + a + 9$.

LIMPING TRIANGLE

A 'limping' triangle is a right triangle (see chapter 5) with the two shorter sides only different in length by 1 unit.

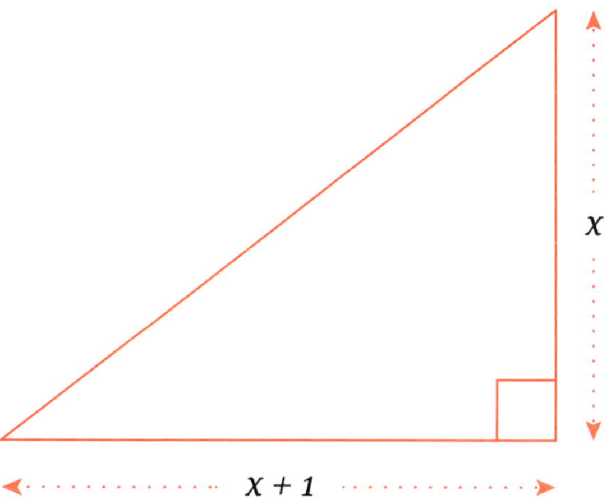

QUIZ QUESTION: There is a limping triangle with hypotenuse (longest side) 29 units long. Find the other two sides.

ANSWER AT THE BACK OF THE BOOK

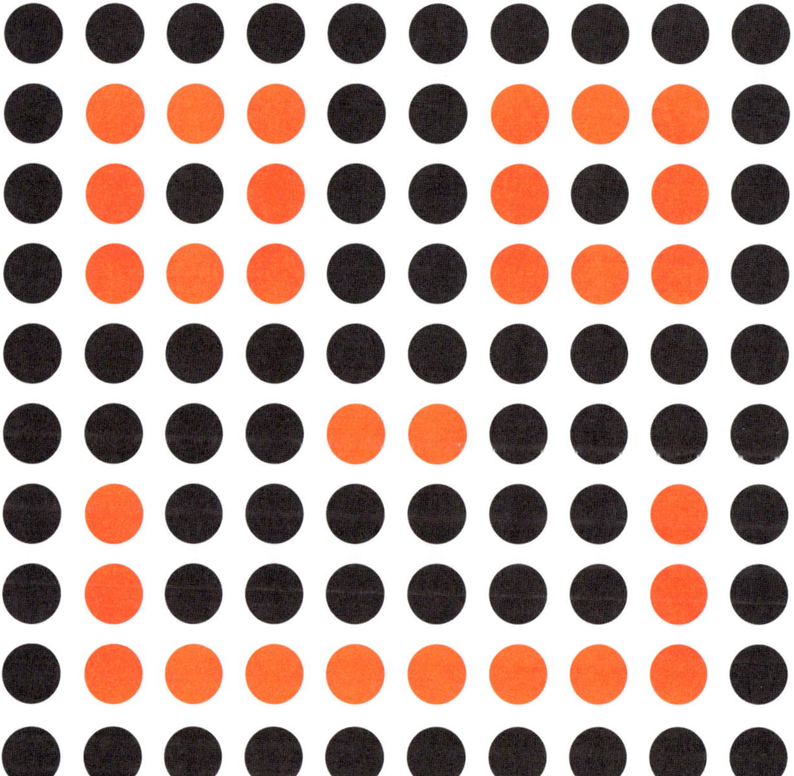

30

DODECAHEDRON

The shape below is called a 'dodecahedron', from the Greek word *dodeka* meaning 'twelve' (*duo* – 2 and *deka* – 10). This Platonic solid has 12 regular pentagonal faces, which meet in groups of three at each vertex.

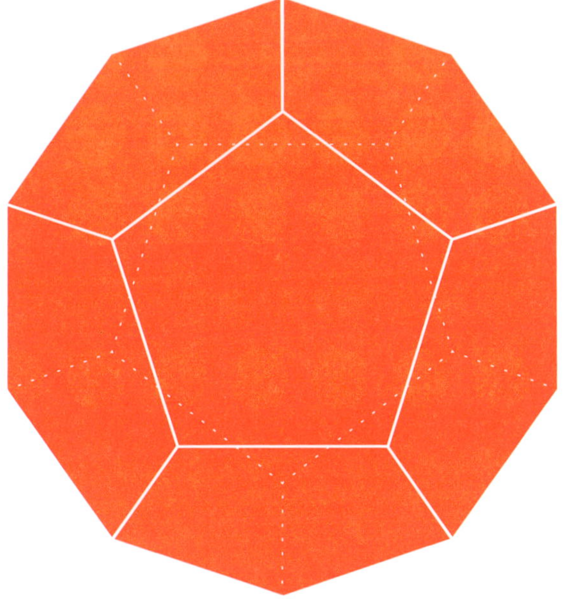

As we know, Leonhard Euler worked out that the number of faces, edges and vertices (corners) of a solid are related to each other by the wonderfully simple equation: $V + F - E = 2$.

So how many edges does this dodecahedron have?

We can see there are 12 faces, each with 5 vertices, so you might think there are 60 vertices. But each vertex is 'attached' to 3 faces so there are really only 20.

Euler's formula for polyhedra tells us $12 + 20 - E = 2$, so there are 30 edges to our dodecahedron.

KING NIELS

of Denmark ruled for 30 years which is impressive in itself. What is amazing is that he did that with a standing army of just 7 men – his personal assistants or *Huskarle* as they were called.

MONTY PYTHON

might've famously and hilariously declared, 'No-one expects the Spanish Inquisition' (trust me, kids, look it up on YouTube), but in fact they *were* expected. The Spanish Inquisition actually gave 30 days notice of anyone whom they planned to 'inquire'.

SLOTHS

do everything slowly – even digest their food. The leaves they eat are complicated for their systems to break down so a sloth can take up to 30 days to digest a meal.

A queen ant can fertilise her eggs with sperm that she has kept in storage for 30 years.

PARTITION PARTY

We've spoken a few times about partitioning numbers (see chapters 4 and 22) and you'll never guess how many partitions there are of the number 9? Okay, sure, it's chapter 30 ... yes ... 9 has 30 partitions.

But if you write a number as the sum of only primes, it is called a 'prime partition'.

For example $9 = 7 + 2$ is a prime partition of 9.

QUIZ QUESTION: **Find all 4 prime partitions of 9.** ANSWER AT THE BACK OF THE BOOK

PRIMORIAL SOUP

Thirty is what's known as 'primorial': 30 = 5#.

I mentioned factorials earlier (see chapter 24). To get a factorial you multiply a number by all the whole numbers less than but including itself.

So 5 factorial or $5! = 5 \times 4 \times 3 \times 2 \times 1 = 120$.

The primorial of any number is the product of all the primes, less than or equal to that number.

So 5 primorial, written 5#, is $5 \times 3 \times 2 = 30$.

Now, 6 isn't prime, so 6# $= 5 \times 3 \times 2 = 30$.

QUIZ QUESTION: 5# = 30, 6# = 30. Find the next three different primorials – by hand, if you're up to the challenge! ANSWER AT THE BACK OF THE BOOK

LUNAR MONTH

A lunar month, the amount of time that it takes the Moon to pass through its phases and orbit the Earth, is close to 30 days (more like 29.5306, if you're fussy) which is why ancient calendars agreed on 30 days as the length of a month. The problem with this was that even the ancients knew that it took about 365 days for the Sun to return to rising in exactly the same position as it did for any given day, for the seasons to repeat, and so on. So the solar year, which is how long it takes the Earth to orbit the Sun – though most of the ancients didn't realise that the Earth went around the Sun (but that's another story altogether) – wasn't an exact number of 30-day months. As a result, some extra days needed to be added. There's a great book by David Ewing Duncan called *Calendar* which goes through the whole saga (see chapter 46 for some more wacky calendar antics).

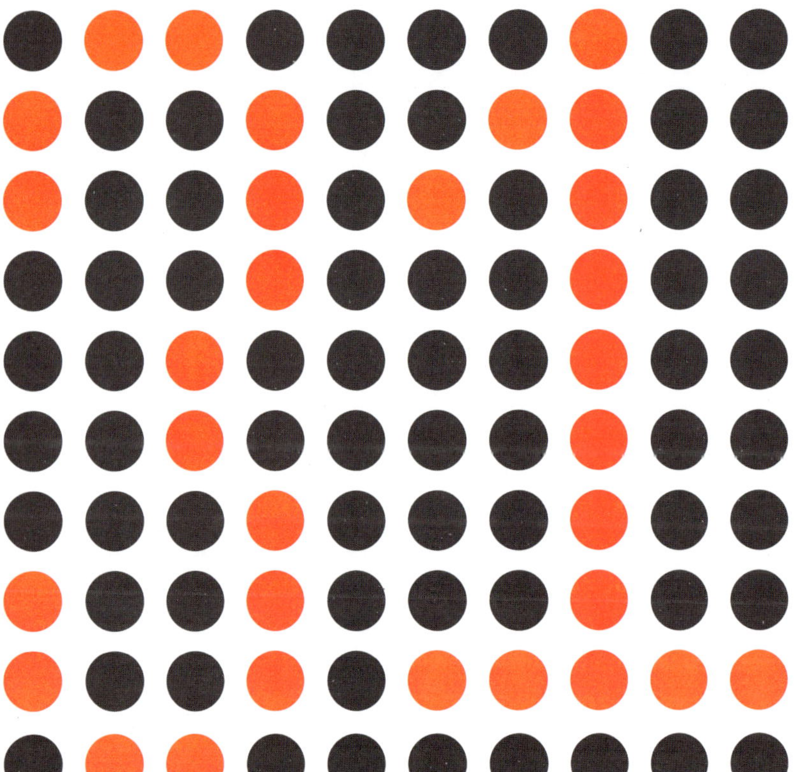

31

THE THIRD MERSENNE PRIME

Because $31 = 2^5 - 1$, it is the third Mersenne prime.

Almost all of the largest prime numbers ever discovered are Mersenne primes (primes equal to $2^p - 1$, where p is prime). The 48th Mersenne prime, discovered in January 2013 by Dr Curtis Cooper, is:

$$2^{57,885,161} - 1$$

It has a whopping 17,425,170 digits and if typed out in the same font as the 7 Harry Potter novels would run half that length again – an incredible 5000 or so pages.

To find out more about these massive primes search the web for the TED talk (http://bit.ly/19RvlHE) given by a dashingly handsome Australian number-loving comedian ... what was his name again ... I've got it lying around somewhere ... that's right ... it's ME!

Initially, the Mersenne primes occur fairly regularly:

$2^2 - 1 = 3$ is the first Mersenne Prime, written as M1,
$2^3 - 1 = 7$ which is M2
$2^5 - 1 = 31 =$ M3
$2^7 - 1 = 127 =$ M4

... but just when you're beginning to think, 'I'm onto something here: for any prime number p, $2^p - 1$ is also prime! Wow ... they'll call this the (insert your name here) theorem ... and I'll be famous!' ... you realise that:

$2^{11} - 1 = 2047 = 23 \times 89$ ... which is *not* prime.
So $2^p - 1$ is not prime for every prime p.

In fact, they keep coming regularly for a while then thin out pretty quickly after that: $2^{13} - 1 = 8{,}191$; $2^{17} - 1 = 131{,}071$ and $2^{19} - 1 = 524{,}287$ are all prime, as is $2^{31} - 1 = 2{,}147{,}483{,}647$ (thanks Euler – again!).

$2^{127} - 1 = 170{,}141{,}183{,}460{,}469{,}231{,}731{,}687{,}303{,}715{,}884{,}105{,}727$ is prime – proven in 1876 with only ink and paper and some mighty brain power. Well done, Mr Édouard Lucas.

Then computers came along and everything changed. From the 1950s on we've found increasingly massive primes ...

$$2^{2203} - 1$$

(October 1952, 664 digits long)

$$2^{11{,}213} - 1$$

(June 1963, 3376 digits long)

$$2^{21{,}701} - 1$$

(October 1978, 6533 digits long)

$$2^{216{,}091} - 1$$

(September 1985, 65,050 digits long)

$$2^{1{,}398{,}269} - 1$$

(November 1996, 420,921 digits long)

$$2^{43{,}112{,}609} - 1$$

(August 2008, 12,978,189 digits long)

... all leading up to Curtis Cooper's 2013 monster.

BOXING WIZARDS!

A 'pangram' is a sentence that contains every letter of an alphabet. It comes from the Greek *pan* meaning 'all' and *gramma* meaning 'thing written, letter of the alphabet'.

Probably the best-known English pangram is 'The quick brown fox jumps over the lazy dog' which contains every letter a to z in only 35 letters.

Changing the second 'the' to 'a' and we get it down to 33 letters.

But we can do better. For a long time, my favourite pangram was the 32-letter 'Pack my box with five dozen liquor jugs'.

But one of the many amazing things I learnt, when reading David Darling's excellent *The Universal Book of Mathematics*, was the pangram 'The five boxing wizards jump quickly', which contains just 31 letters!

Here are a few more pangrams – but to be honest a couple are a little bit contrived ...

(50) We promptly judged antique ivory buckles for the next prize.

(48) Sixty zippers were quickly picked from the woven jute bag.

(46) Crazy Fredrick bought many very exquisite opal jewels.

(32) Pack my box with five dozen liquor jugs.

(31) The five boxing wizards jump quickly.

(30) How quickly daft jumping zebras vex.

(29) Sphinx of black quartz, judge my vow.

(28) Waltz, nymph, for quick jigs vex bud.

32

The world's shortest national anthem (by text length) is *Kimigayo*, the Japanese anthem, at 11 bars long and comprising a mere 32 *kanji* (Japanese alphabetical) characters. But what it lacks in length, it makes up for in poetry – the English translation is:

君が代は	*May your reign continue for*
千代に八千代に	*a thousand, eight thousand, generations,*
さざれ（細）石の	*Until the pebbles*
いわお（巌）となりて	*Grow into boulders*
こけ（苔）の生すまで	*Lush with moss*

It also happens to be one of the world's oldest national anthems.

MAGIC HEXAGRAM PART 1

In his delightful book, *Are Universes Thicker Than Blackberries?*, the American Martin Gardner, one of the all-time great popularisers of mathematics, explains the creation of a magic hexagram.

In this hexagram we have placed the numbers 1 to 12 so that each row of 5 triangles adds up to 32.

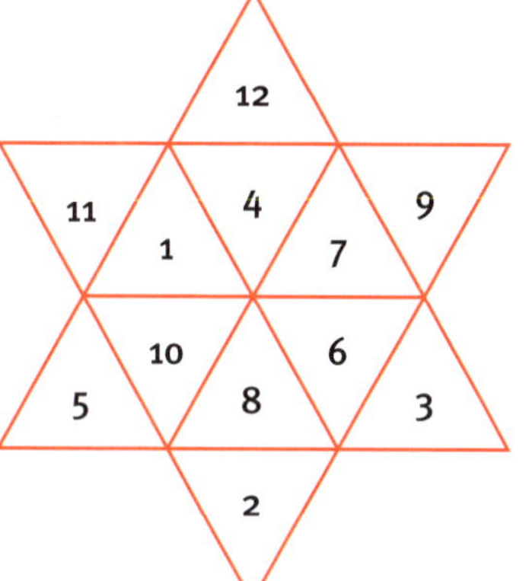

In fact, it is the only magic hexagram with constant 32, and one of only two such hexagrams using the numbers 1 to 12.

TESSERACT

We all know what a square looks like and that a cube is the 3-dimensional version (a mathematician would say 'generalisation' or 'analogue') of a square.

Indeed, we can picture a cube in our mind by taking 6 squares in 2 dimensions, like the diagram on the right and by joining them up in 3 dimensions by holding one square in the same place, 'gluing' all the sides together. This gives us our 6 faces, 8 vertices (corners) and 12 edges that make a single cube.

All we've really done is join the point A to B by gluing two sides together and then we've glued up the sides indicated by arrows.

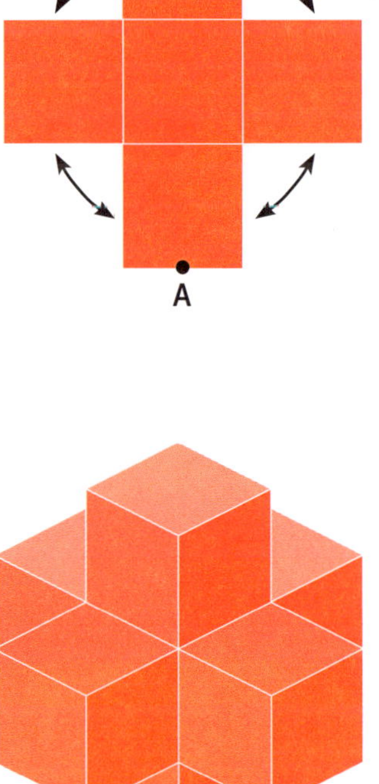

As we saw in chapter 24, a 'tesseract' is the 4-dimensional version of a cube. It has 32 edges and 16 vertices and contains 24 squares and 8 cubes.

To get some of the idea of the geometry of a tesseract, we take 8 cubes in 3 dimensions – a line of 4 cubes, with one coming out in each direction from the second cube in the line.

Now, in the same way we glued up the squares to get a cube, we glue up the cubes to get a tesseract. In 4 dimensions we pull the bottom cube around to the top of the line of cubes, and glue these two faces together – then glue up the other faces of the cubes coming out in all the directions.

Think about it for a few seconds. Ow!

Don't worry, right about now my head hurts too.

Leeches are amazing animals.

In medicine, leeches have been used for everything from removing blood from people, to helping reattach body parts and assist blood flow and recovery from certain types of surgery. They are hermaphrodites and have up to 9 pairs of testes. And while it is sometimes said that they have 32 brains, more accurately their bodies come in 32 segments and they have one brain which comes in 32 ganglia: one in each segment of their bodies. Impressive, nonetheless.

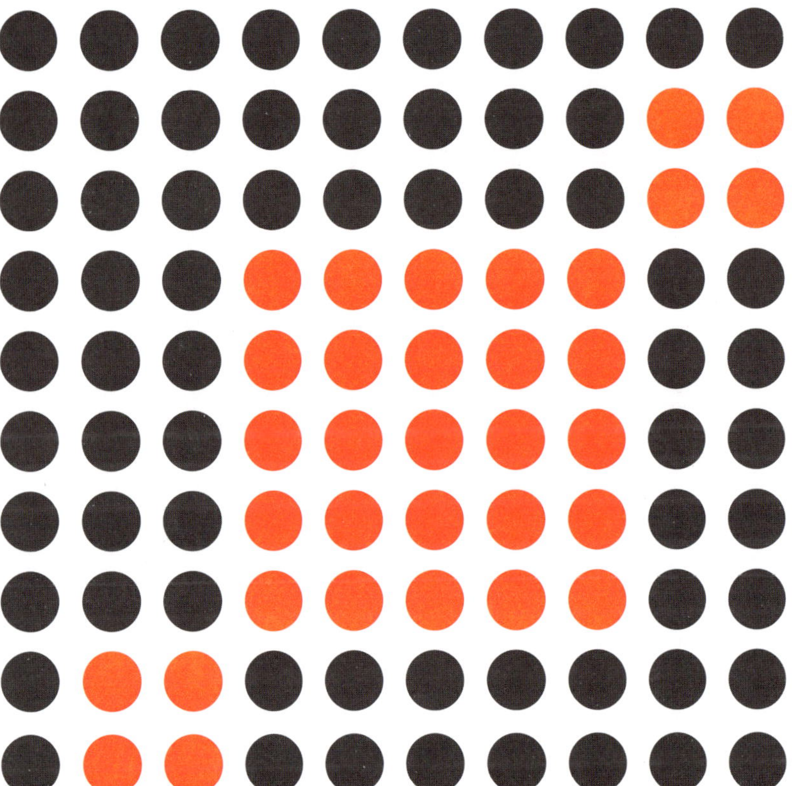

33

SSHHH!

The highest rank or degree one can attain in the Ancient and Accepted Scottish Rite of Freemasonry is the 33rd degree. You must get to do the most awesome secret handshake by that stage!

BIZARRO

Here's a bizarre-looking equation:

$33 = 1! + 2! + 3! + 4!$

See chapter 6 if you don't know what I'm going on about.

TRIANGULAR

The largest number that cannot be written as the sum of distinct triangular numbers is 33.

QUIZ QUESTION: Write the following numbers as the sum of distinct triangular numbers: 55, 64, 90. ANSWER AT THE BACK OF THE BOOK

BOTTLES OF 33 BEER ON THE WALL

If you're thirsting for a beer, don't just walk into a bar and ask for a '33'. It could mean many different things. Because 330 mls is roughly a third of a litre and therefore a popular bottle size, depending where you are in the world you could be served any of various '33 Export Lagers' which are found everywhere from Vietnam to France to Nigeria; something from 33 Acres Brewing Co in Vancouver; or in the US quite possibly a Rolling Rock pale lager, which while being called 'Rolling Rock' still has a '33' displayed prominently on its label. The reason is that when the beer's 'pledge of quality' was scripted, someone wanted to show how short it was by pointing out it contained only 33 words. So they wrote a great big 33 on the page. It stuck.

MAGIC HEXAGRAM PART 2

We just met this magic hexagram (see chapter 32):

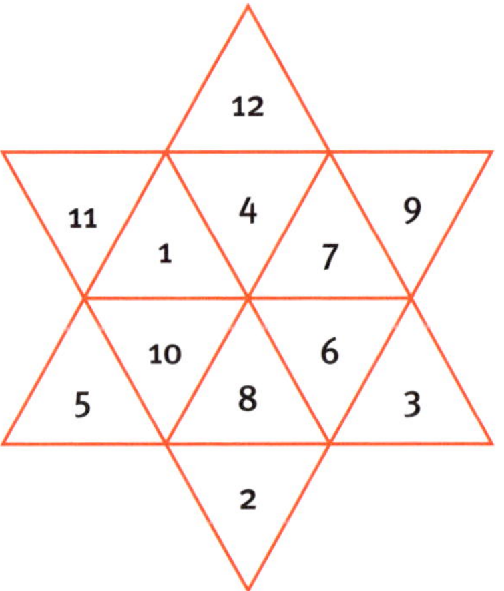

But now we're going to turn this hexagram into another one. Go through triangle by triangle and each time replace the number by subtracting the current number from 13. So the 12 up the top becomes 13 − 12 = 1, the 11 becomes 13 − 11 = 2, the 1 becomes 12 and so on.

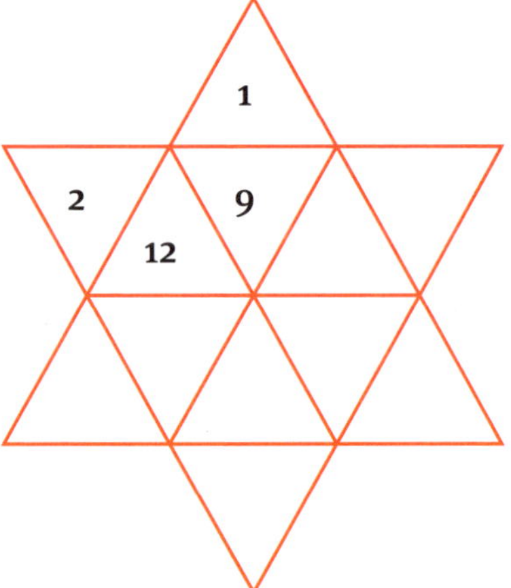

QUIZ QUESTION: What do you notice about this new hexagram?

ANSWER AT THE BACK OF THE BOOK

MOROŃ'S RECTANGLE

We saw in chapter 21 the amazing result by A. J. W. Duijvestijn that a square can be broken in 21 smaller squares of different sizes. Over 50 years before that discovery, Polish mathematics whiz Zbigniew Moroń came very close when he discovered this 33×32 rectangle that can be composed from 9 unequal squares.

If you look at the square in chapter 21 and this rectangle we can turn the shapes into equations. Moroń's rectangle shows us that:

$$1^2 + 4^2 + 7^2 + 8^2 + 9^2 + 10^2 + 14^2 + 15^2 + 18^2 = 33 \times 32 = 1056$$

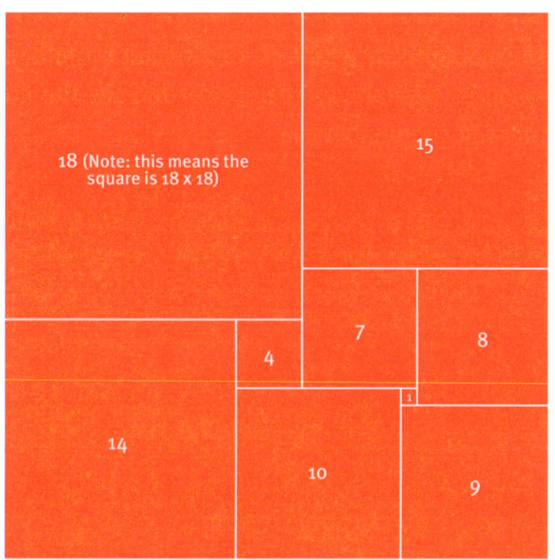

... and AJW's square illustrates the rather longer equation:

$$2^2 + 4^2 + 6^2 + 7^2 + 8^2 + 9^2 + 11^2 + 15^2 + 16^2 + 17^2 + 18^2 + 19^2 + 24^2 + 25^2 + 27^2 + 29^2 + 33^2 + 35^2 + 37^2 + 42^2 + 50^2 = 112 \times 112 = 12{,}544$$

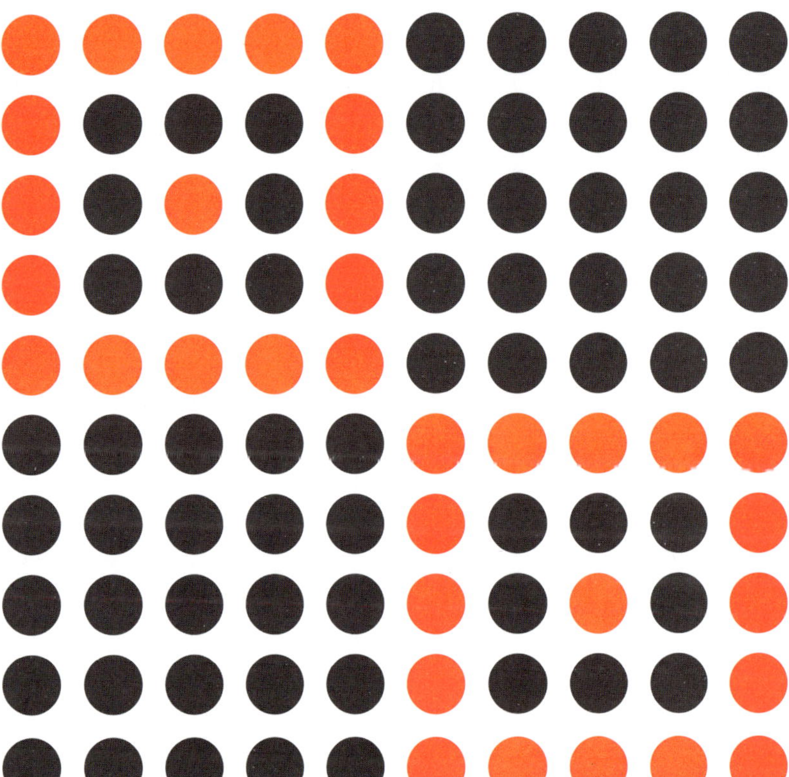

34

In 2015, French was officially spoken in 34 countries around the world. Unfortunately for the Francophiles among you, exactly twice as many – 68 – spoke English. *C'est la vie.*

MAGIC SQUARES

In a 4 × 4 magic square, the numbers 1 to 16 occur once each, and all the rows, columns and diagonals add up to 34. The 3 × 3 magic square we saw in chapter 9 is essentially the only 3 × 3 magic square. But there are 880 different 4 × 4 ones! If you think that's a lot, make sure you're sitting down when you get to chapter 65 ...

QUIZ QUESTION: Try finishing off these 4 x 4 magic squares.

ANSWER AT THE BACK OF THE BOOK

12	13		8
6			10
		14	11
9	16		5

1	2		16
13		3	
	7		5
8		6	

10	16	1	7
15	5	12	2

16			13
		11	
4			1

MORE MAGIC

Here's another square that seems to be even more magic – until you realise what's going on. Check out the first table on the right.

1	2	3	4
5	6	7	8
9	10	11	12
13	14	15	16

Choose any 4 numbers from this grid on the proviso that no two numbers are in the same row or column and add up those 4 numbers.

So maybe you chose:

1 + 6 + 12 + 15 = 34

Or 2 + 5 + 11 + 16 = 34

Or 4 + 5 + 11 + 14 = 34

Or ... whichever 4 numbers you chose you'll get 34. Wow!

If you add this outside column and row to the grid you'll see what's going on.

	1	2	3	4
0	1	2	3	4
4	5	6	7	8
8	9	10	11	12
12	13	14	15	16

Every number in the original grid is just the sum of the number of the row and the column in which it appears – for example 11 = 3 + 8.

So when you add up a number from each row and column you're just adding 1 + 2 + 3 + 4 + 0 + 4 + 8 + 12 which always equals 34.

What's even cooler is you can actually jumble the (1, 2, 3, 4) and (0, 4, 8, 12) and get a new square that seems completely random, but still does exactly the same.

	1	3	4	2
8	9	11	12	10
12	13	15	16	14
4	5	7	8	6
0	1	3	4	2

So, as you can see in this third grid on the right, it also obeys the '34 rule'.

HYPOTRACEABLE DIAGRAMS

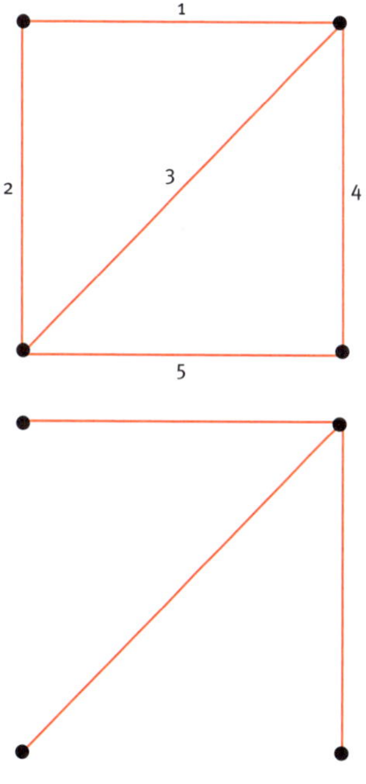

Mathematicians call a path like the one on the right 'traceable' because you can visit each point once and once only without lifting your pencil off the page.

For example, place your pencil in the top right hand corner and trace line 1, then 2, then 3. Or, starting in the top left hand corner, take the path 2, 3, 4. From the bottom left, trace 3, 4, 5, (or 5, 4, 1) ... and so on.

You should be able to see that the second diagram on the right is 'untraceable'.

Now, the path below is also untraceable. It's what is called the Thomassen Graph 34.

Now, here's an excellent exercise – remove any point and its connecting edges from the diagram and the redrawn diagram is traceable. We call this a hypotraceable graph.

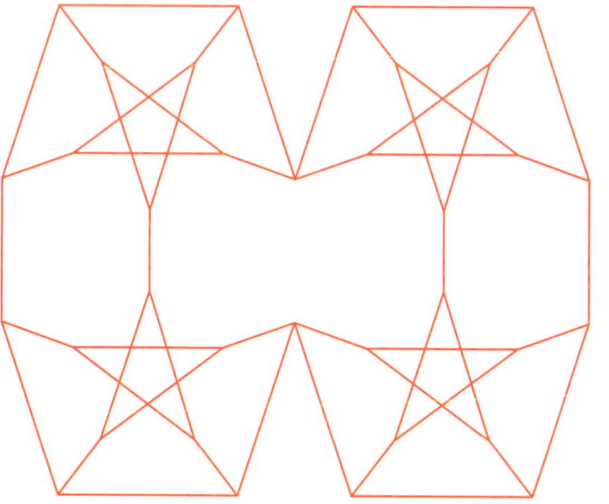

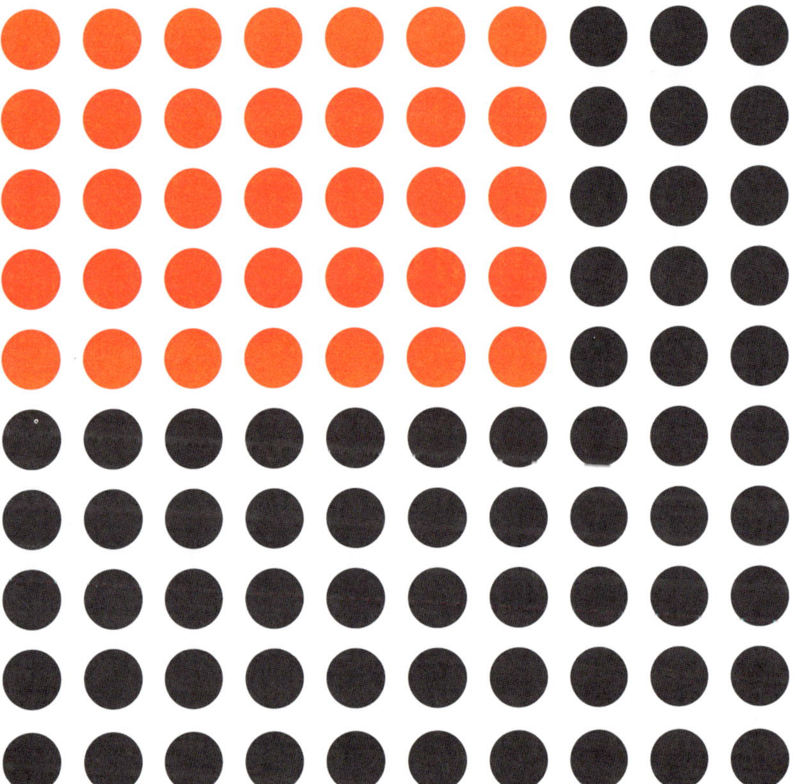

35

The world's first scientific pocket calculator was the HP-35 so named because it had 35 keys and set you back a cool $400 ... which is over $2000 today.

APPROXIMATION OF PI

This very simple diagram – a square inside a circle inside a bigger square – is actually the first step to trying to calculate π.

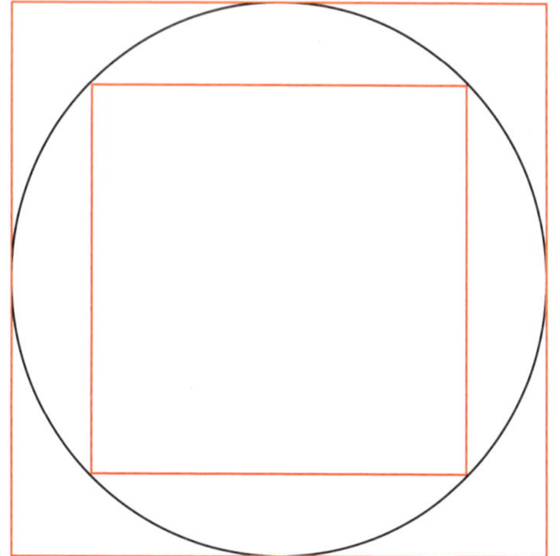

We've spoken about π (see chapter 22), the distance around a circle that is 1 unit across.

Let's say this circle is 1 unit across.

That means the smaller square has a diagonal of length 1 and the larger square has an edge of length 1.

Clearly the distance around the circle is in between the distances around the two squares.

This observation tells us that:

$2\sqrt{2} < \pi < 4$.

It's not a great approximation, in fact, it only tells us that π is somewhere between 2.83 and 4.

But if we 'inscribe' a circle inside two shapes that have more than 4 sides, you'll see that as there is less left over space between the boundaries of the shapes the approximation of π gets better.

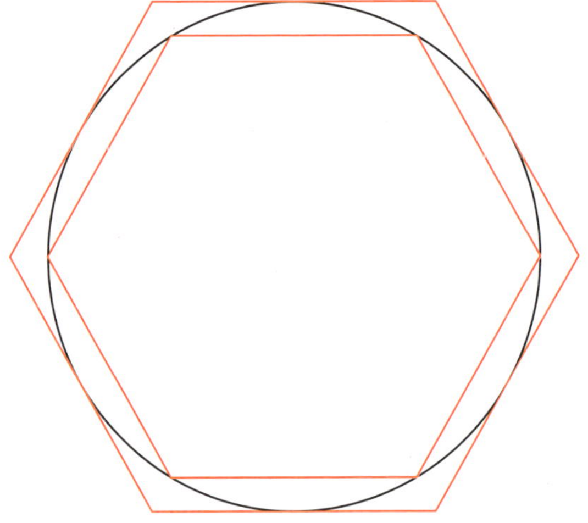

Using a hexagon we get $3 < \pi < 3.464$ and so on with shapes with more sides.

I for one have nothing but admiration for someone who can dedicate a significant part of their life to a single obsession. Enter Dutch mathematician and fencing coach Ludolph van Ceulen who, after years of calculations, deduced that using a shape with 2^{62} sides (that's 4,611,686,018,427,387,904 sides – he couldn't draw it but he could work out how far it was around the perimeter) ...

we get $\pi = 3.14159265358979323846264338327950288$ ...

That's π to 35 decimal places and that's good enough for me! Well done, Ludolph. When they laid him to rest in 1610, they inscribed the so-called 'Ludolphine' number on his tombstone.

In chess, a knight moves 2 squares forward or to the side, then 1 square around the corner. So the knight on the middle square could move to any of these 8 positions:

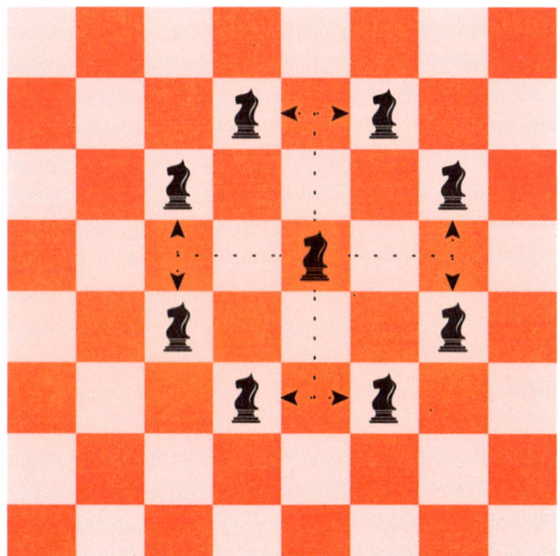

On a board 8 squares long and 8 squares wide, a knight can travel 35 steps without crossing back over its path. See if you can find the 35-step path.

HEX-ACTLY

A polyomino (not as scary as it sounds) with 6 squares is called a 'hexomino'. A fun way to spend an afternoon, if you don't have anything else planned, is to try and find all 35 hexominoes. Here are 4 to get you started:

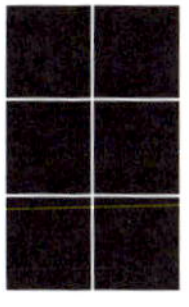  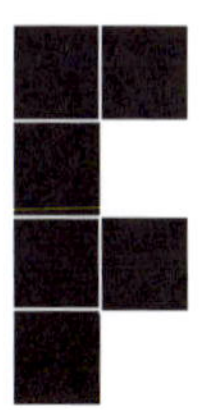 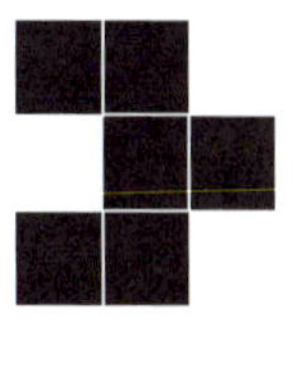

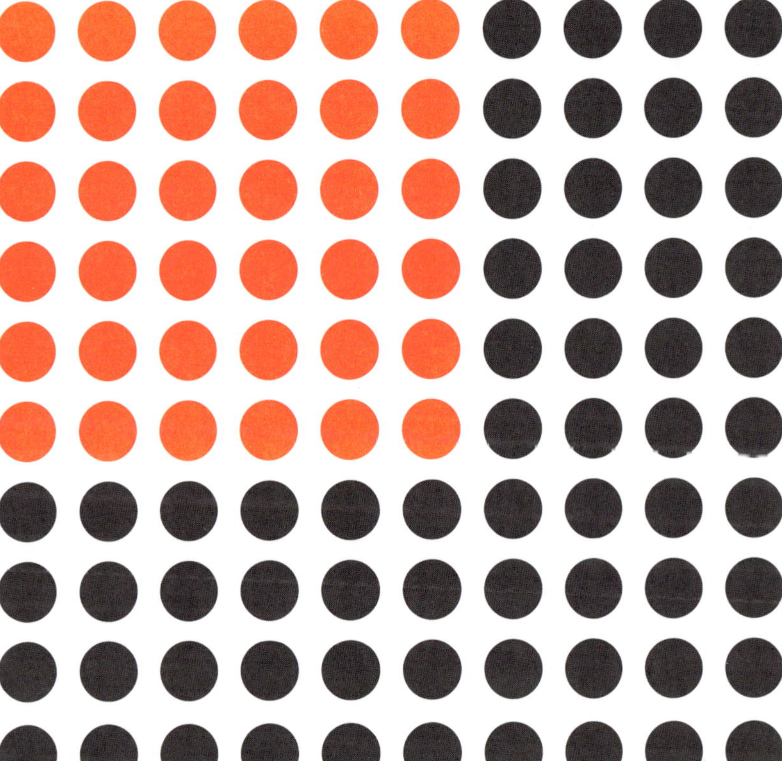

36

PENROSE TILING

One of the grooviest mathematicians of recent times (I'd actually say ever, but I've met the guy so I'm a little bit biased) is the Englishman Sir Roger Penrose. He has done incredible work in the fields of cosmology, quantum mechanics and popular mathematics.

And one of the most beautiful things he has discovered is the 'Penrose Tiling of the Plane'.

Imagine that you were rich ... like *really* rich ... so rich that your bathroom wasn't just massive, it was infinitely large. That's right, your bathroom has a wall that goes forever both in length and height.

Pretty cool. But how would you choose your bathroom tiles!

Look, it's a silly question, but the mathematics of tiling an infinite plane are amazing.

Penrose discovered two tiles, based on the angle 36 degrees:

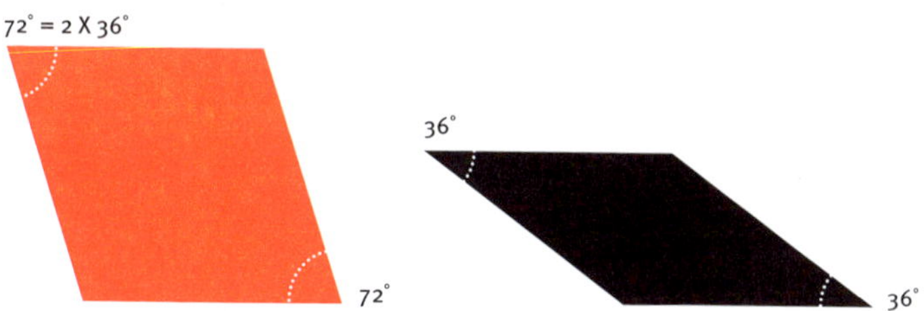

... that can be arranged to tile the plane 'aperiodically'. That means that you can slide this pattern anywhere you want and it never covers itself perfectly. Opposite is one we prepared earlier.

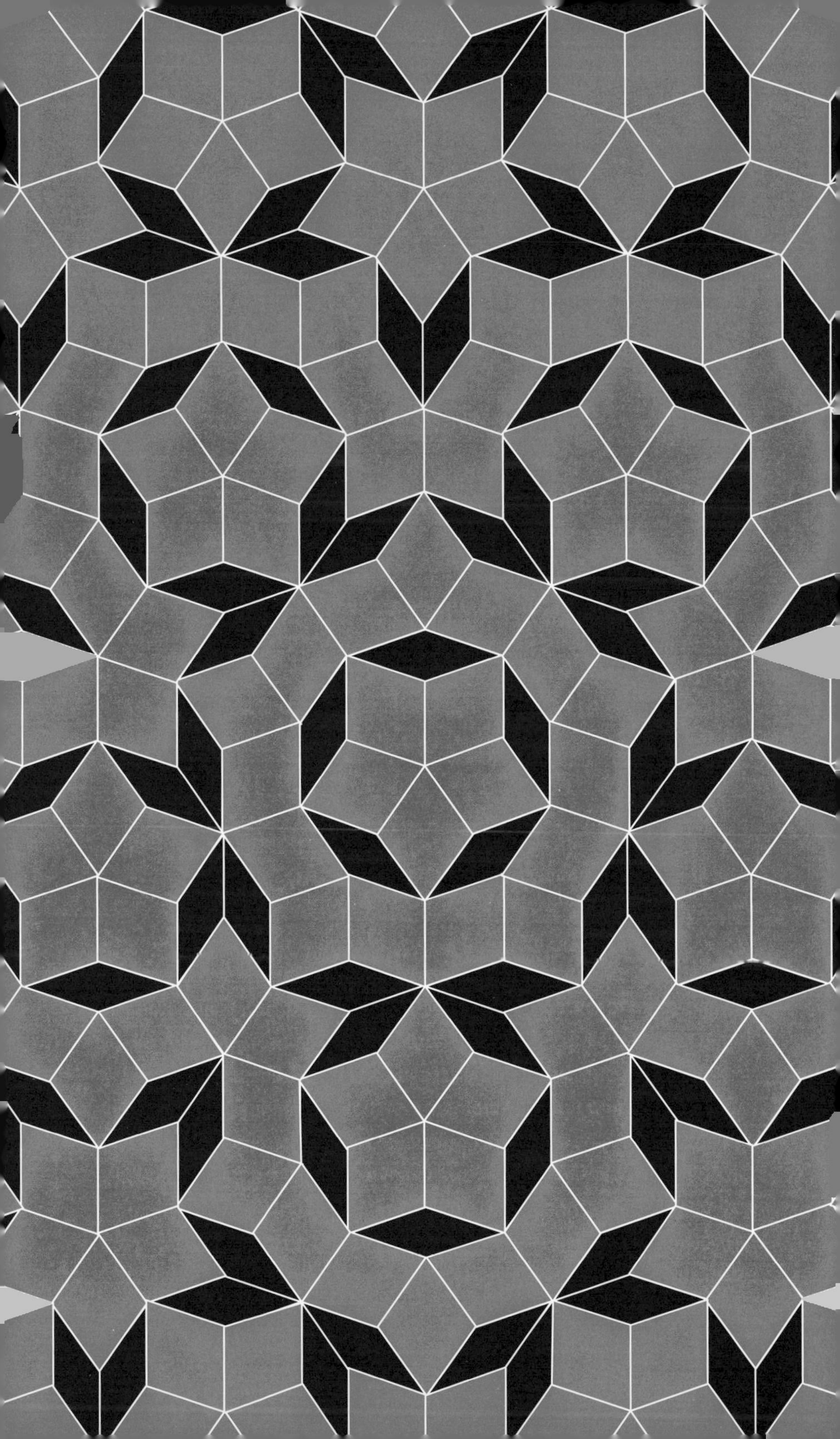

A number is composite if it has divisors other than 1 and itself. So 14 is composite because $14 = 2 \times 7$, but you can easily check (by a bit of division) that 17 can only be written as 17×1. Seventeen isn't composite; it's a prime number. Thirty-six, on the other hand, is not only composite, it's *highly* composite. This was a concept invented by the brilliant Indian mathematician Ramanujan for numbers that have more divisors than any other below them. So 36 has divisors 1, 2, 3, 4, 6, 9, 12, 18 and 36 and no number from 1 to 35 has 9 divisors. Other highly composite numbers include 332,640; 43,243,200 and 6,746,328,388,800 (which was the highest of the 102 composite numbers calculated by Ramanujan in 1915).

QUIZ QUESTION: There are actually 9 highly composite numbers less than 100. Find all of them. ANSWER AT THE BACK OF THE BOOK

TWO DICE

If you roll two dice, they can each come up any of 6 possible ways. So there are 36 possible rolls. Obviously, there aren't 36 different totals that you could get. This is because different rolls can give the same total. So a 1 and a 4 give a total of 5, but so do a 2 and a 3 or a 3 and a 2 or a 4 and a 1. The table shows all the possible totals:

First Roll

Second Roll	1	2	3	4	5	6
1	2	3	4	5	6	7
2	3	4	5	6	7	8
3	4	5	6	7	8	9
4	5	6	7	8	9	10
5	6	7	8	9	10	11
6	7	8	9	10	11	12

The most likely result is for two dice to total 7 and this will happen on average $\frac{6}{36}$ or $\frac{1}{6}$ of the time.

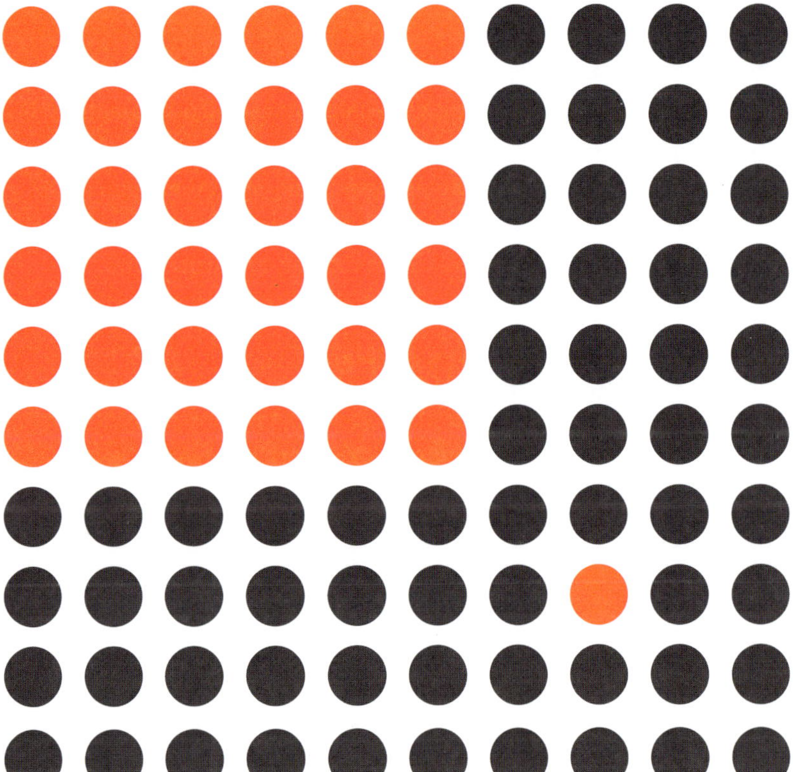

37

An attempt in 1879 in Liege, Belgium to revolutionise their delivery of mail by using 37 trained cats ... failed abysmally.

The Australian Rules football legend, and 2014 Australian of the Year, Adam Goodes, (disclaimer: my mate) wore the number 37 for the mighty Sydney Swans (disclaimer: my favourite team).

HEXAGONAL NUMBERS

Thirty-seven is the fourth 'centred hexagonal' number. We get these by placing hexagons around a centre. Not surprising, really. The centred hexagonal numbers are 1, 7, 19, 37 and so on.

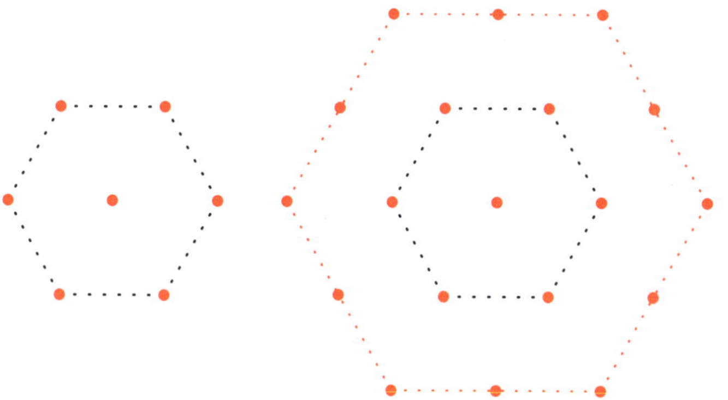

BRIDGE

I've never played Bridge (I know, what sort of a nerd do I call myself – I'm more of a chess guy, to be honest) but I'm reliably informed that 37 is the most points possible in a Bridge hand (4 aces at 4 points each, 4 kings (3), 4 queens (2) and a jack).

THE SECRETARY'S PROBLEM

The 'secretary's problem', also known as the 'sultan's dowry' and by many other names, is a particularly famous mathematical puzzle.

Imagine you are going to meet 100 people, one at a time, and each of them will offer you an amount of money. You've got 30 seconds to accept the amount of money or to tell the person, 'No thanks, please go away'. At some stage you will accept an offer, take that amount and not meet anyone else in the line of 100 people. But after you have told someone to go away, you can't ask them to come back later on and take the money they offered.

To win the game, you have to say yes to the person with the most money in the line.

Ask yourself this ... how many of the 100 people will I allow myself to meet to give myself a feel for what's on offer, before I decide, 'Okay – this is it. I'll say yes to the next person I meet who is beating all the previous offers'?

Do you just meet 2 or 3 people and then dive on in at the next 'big offer' you get? Or do you wait for 85 or so people so that you have learnt a lot about the potential offers, but risk the chance that the best offer has already passed you by? How many offers should you reject before you think, 'Okay, I'm ready'?

Thousands of mathematicians have wrestled with this problem using lots of different mathematical methods. The agreed best strategy is to let 37 of the offers go, then jump in the next time you get 'the best offer yet'.

If you do this, you also have a 37% chance of winning the game – surprisingly good odds, if you ask me.

The number 37 arises as $\frac{1}{e}$ where e is a very important irrational number: e = 2.718 ... so $\frac{1}{e}$ is about 0.37 or 37%.

The number 37 is written in the heavens – the open cluster NGC2169, found 3,600 light years from here towards the constellation Orion, actually looks like a 37.

PICK A 2-DIGIT NUMBER ...

Now, multiply it by 3. Multiply that by 7. Multiply that by 13.

Multiply that by 37. What have you got?

Wow! Why?

Well, $3 \times 7 \times 13 \times 37 = 10{,}101$ and when you multiply a 2-digit number by 10,101 the ones make the 2-digit number repeat and the zeros spread it out nicely.

If I told you that $1{,}001{,}001 = 3 \times 333{,}667$ could you create a version of this puzzle that works for 3-digit numbers?

Another cool thing about 37: if a 3-digit number *abc* is divisible by 37 so is the number *bca*. For example, 37 divides into 851 ($851 = 37 \times 23$) so we know straightaway that 37 divides into 518.

This all comes down to the fact that $999 = 37 \times 27$.

This is true because the number written with digits abc is really the number $100a + 10b + c$ that is, $851 = 100 \times 8 + 10 \times 5 + 1$.

Say we have two numbers A and B, where A is written *abc* and B is has the digits *bca*:

What is $10A - B$?

$10A - B = 10\,(100a + 10b + c) - (100b + 10c + a)$

$= 1{,}000a + 100b + 10c - 100b - 10c - a$

$= 999a$

Now, 999 is divisible by 37 so, $10A - B$ must be divisible by 37.

So if A is divisible by 37, it should be obvious that B is also divisible by 37. Simple ;)

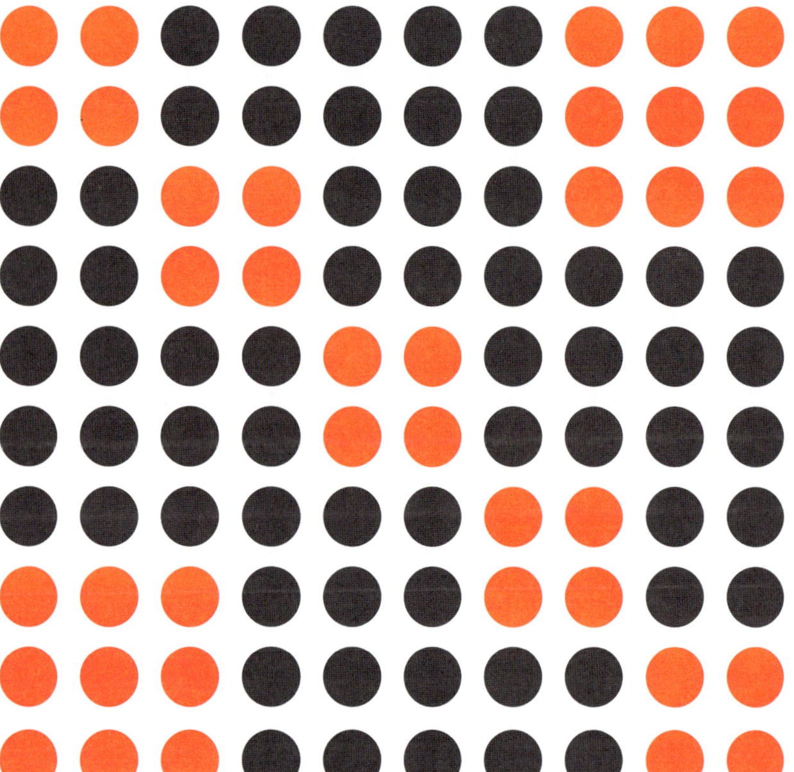

38

BIG BALLS

In 2000, table tennis balls were enlarged from 38 mm to 40 mm.

HUH?

Spot the continuity error: in *Raiders of the Lost Ark* during the firefight in the bar, our hero Indiana Jones starts out with a .38 revolver – which changes to a .45, back to a .38, and back once again to a .45.

CREME PUFF

According to the *Guinness Book of World Records* the oldest cat to ever live was Creme Puff, born on 3 August 1967, who lived an amazing 38 years and 3 days. Creme Puff lived with her owner, Jake Perry, in Austin, Texas, US. Perry was also the owner of Grandpa Rex Allen, the previous holder of the title world's oldest cat.

MAGIC ~~SQUARES~~ HEXAGONS

Remember magic squares? (Check out chapters 9 or 34 if you can't.) Well, there is only one magic hexagon. It uses the numbers 1 through to 19, with every straight line of three, four or five hexagons adding up to 38. Try and complete it. Answer, of course, at the back of the book.

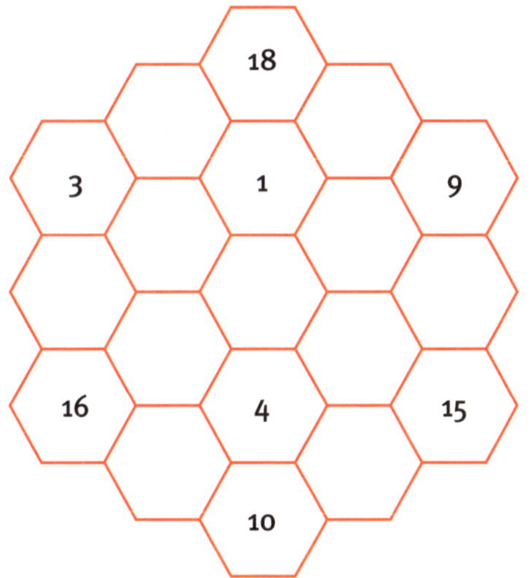

American Scott Flansburg is known as the world's fastest human calculator.

On 27 April 2000 on the set of *Guinness Book of World Records* in Wembley, UK, he correctly added a randomly selected number (38) to itself 36 times in 15 seconds without using a calculator!

While most people have heard of Halley's Comet (see chapter 76), fewer know of the Stephan–Oterma Comet which quietly goes about its business orbiting from near Mars all the way out to Uranus and back every 38 years.

DOUBLE YOUR (UN)LUCK

The simple addition of the 00 to a roulette wheel takes the number of slots from 37 to 38 and almost doubles the house's chances of winning from 2.7% to 5.26%.

FACTORIAL PRIME TIME

You remember factorials from chapters 6 and 24? Well, look at this:

$$3! - 1 = 3 \times 2 \times 1 - 1 = 5$$
$$3! + 1 = 3 \times 2 \times 1 + 1 = 7$$
$$4! - 1 = 4 \times 3 \times 2 \times 1 - 1 = 23$$

Now, the numbers 5, 7 and 23 are prime. But just when we think we might be noticing something pretty cool about all numbers of the form $n! + 1$ and $n! - 1$ being prime, along comes

$4! + 1 = 25$ to blow that little idea out of the water.

When $n! - 1$ or $n! + 1$ is a prime, we call it a 'factorial prime' and they're pretty rare. We only know of 26 factorial primes of the form $n! - 1$ and 21 of the form $n! + 1$. We'll see them again in chapter 77.

We do know that 38! – 1 = 523,022,617,466,601,111,760,007,224,100,074,291,199,999,999 is the 16th factorial prime.

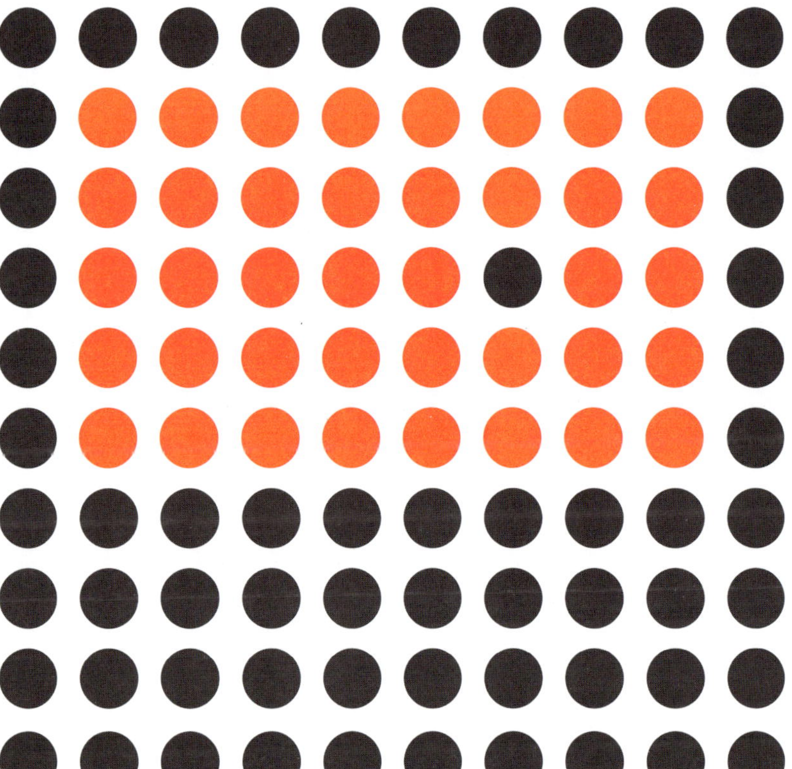

39

The number 10^{39}, or 1 with 39 zeros after it, is called a 'duodecillion'. If you suspect there may be one hanging around your house, it looks a bit like this: 1,000,000,000,000,000,000,000,000,000,000,000,000,000.

UNINTERESTING 39

Thirty-nine has been described as the first 'uninteresting' number. If you think about it, if that's true, doesn't being the first uninteresting number in itself make 39 interesting? Ow – paradox alert!

While it's not as sexy as 7 or as nifty as 99, it *has* got a few things going for it. For example, $39 = 3 \times 9 + 3 + 9$. Not impressed? Well, this pattern holds for *any* number ending in 9. Why? It's simple.

A number with digits ab is equal to $10a + b$. So here, $39 = 3 \times 10 + 9$

The example $39 = 3 \times 9 + 3 + 9$ matches up $10a + b$ and $a \times b + a + b$. Let's solve that $10a + b = a \times b + a + b$

Subtract b from each side and you get $10a = a \times b + a$

Subtract a from each side and you get $9a = a \times b$

As long as a isn't a zero, divide each side by a and get $b = 9$. So:

$39 = 3 \times 9 + 3 + 9$

$79 = 7 \times 9 + 7 + 9$

$159 = 15 \times 9 + 15 + 9$ and so on.

PARTITION PROBLEM

One partition of 39 is $39 = 4 + 15 + 20$

And if you multiply those terms you get $4 \times 15 \times 20 = 1200$.

QUIZ QUESTION: **There are two other 3-number partitions of 39 that also multiply to 1200. Find them.** HINT: In any potential partition, start with the smallest number and work out what the product of the other two would have to give you. So if a partition has 4 in it, we need two numbers that add to give 35 and multiply to give $1200 / 4 = 300$. We get 15 and 20. Try this for other small numbers (less than 10). ANSWER AT THE BACK OF THE BOOK

NARCISSISTIC 39

Look at the following three equations and convince yourself that they are correct:

$153 = 1^3 + 5^3 + 3^3$

$371 = 3^3 + 7^3 + 1^3$

$9474 = 9^4 + 4^4 + 7^4 + 4^4$

They all match the description of an n-digit number that is equal to the sum of nth power of its digits. We call 153, 371 and 9474 'narcissistic' numbers.

QUIZ QUESTION: Which of the following numbers are narcissistic? 205, 407, 1546, 8208, 93084, 223443, 1741725, 555555555, 4679307774 ANSWER AT THE BACK OF THE BOOK

But why is this entry under the number 39? Well, brace yourself because the largest narcissistic number is:

115,132,219,018,763,992,565,095,597,973,971,522,401 which has a whopping 39 digits and is equal to ... here we go ...

$1^{39} + 1^{39} + 5^{39} + 1^{39} + 3^{39} + 2^{39} + 2^{39} + 1^{39} + \ldots + 4^{39} + 0^{39} + 1^{39}$

Okay, I definitely need a lie down now.

RECIPROCALS REVISITED

In chapter 17 we met the 'reciprocal' of 17 and its 16-digit repetend:

$\frac{1}{17}$ = 0.0588235294117647 0588235294117647 0588235294117647 ...

Check out the reciprocal of 7:

$\frac{1}{7}$ = 0.142857 142857 142857 with the 6-digit repetend 142857

You might see these two reciprocals and start thinking that any time we have a prime number p, the decimal expansion of $\frac{1}{p}$ will have a repetend $(p - 1)$ units long – it certainly happened for $p = 7$ and $p = 17$.

Turns out it's not always true, but it's close. When p is prime, $\frac{1}{p}$ actually has a repetend that is 'a factor of $(p - 1)$' digits long.

So for $\frac{1}{13}$ our repetend will have length that is a factor of 12.

Checking, we see that:

$\frac{1}{13}$ = 0.076923 076923 076923 ... a repetend of length 6, with 6 being a factor of 12.

If you're following all this you're doing really well. This is beautiful stuff but a little bit complicated.

For reasons I won't go into here, when p and q are prime and both not equal to 2 or 5 we know $\frac{1}{pq}$ will be a repeating decimal.

Now $39 = 3 \times 13 = p \times q$ where p and q are prime and both not equal to 2 or 5 and yes:

$\frac{1}{39}$ = 0.025641 025641 025641 ...

Just gorgeous, as far as I'm concerned.

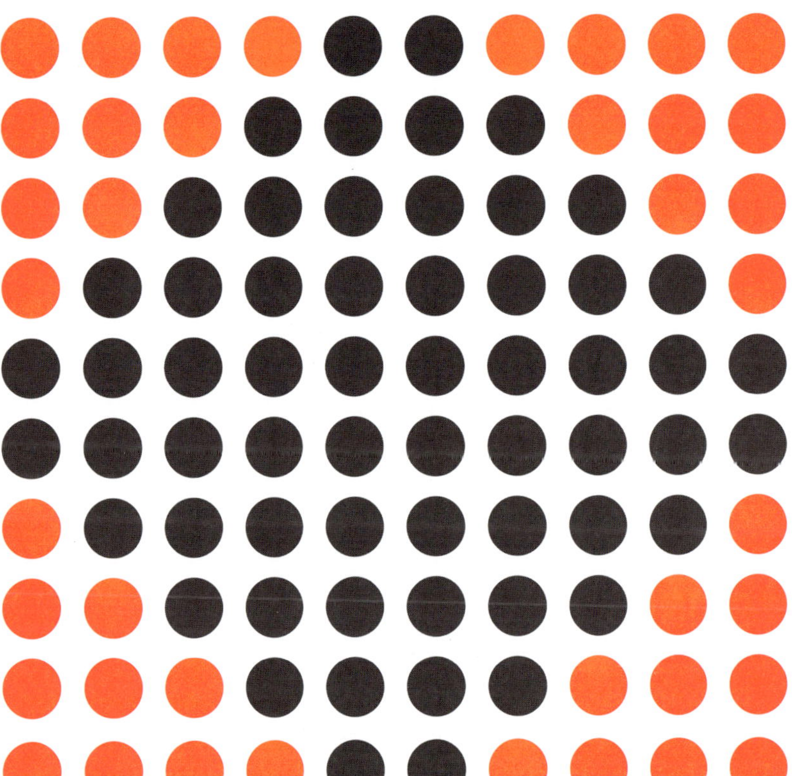

40

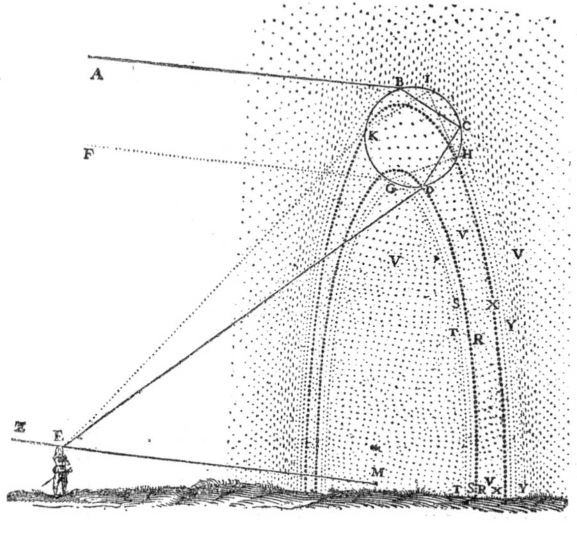

Descartes' sketch illustrating how rainbows are formed. Source: Wikipedia

RAINBOWS

The great French mathematician René Descartes understood how we see a rainbow way back in the 1630s. He noticed that rainbows don't just occur in the sky, they can happen near a water fountain, or when the Sun hits a pane of glass and so on. He took a large sphere of glass, looked through it to the Sun and noted that as he moved the sphere the colours that he saw changed.

He realised when you look at a raindrop in front of you, with the Sun behind you and the angle from the Sun to the raindrop to your eye just right, the full spectrum of light passes into the drop of water, refracts (bends), reflects off the back of the raindrop and refracts again as it leaves the drop and travels towards your eyes.

But when the light is 'refracted', the amount it is refracted depends on its wavelength and hence its colour. This has the effect of 'spreading out' the light into its range of colours. The angle between the Sun, raindrop and your eye will determine which colour you see from that raindrop.

The collection of thousands of raindrops reflect the range of colours back to you and you see these in bands which form the rainbow.

What this means is that if you and a friend are standing a few metres apart looking at a rainbow, strictly you're looking at two different rainbows because a single raindrop that sends a certain colour to your eye will send a slightly different colour to your friend.

That crucial angle is between 40.6 and 42 degrees.

BACHET'S PROBLEM

What are the four weights that can be used on a scale pan to weigh any whole number of grams from 1 to 40 inclusive, if the weights can be placed in either side of the scale pans? This is Bachet's famous problem from 1612. The answer is 1, 3, 9 and 27. To work it out, you have to notice that a scale can weigh something directly.

But you can also weigh something by putting weights on both sides.

It's good fun, in a really geeky way, to convince yourself that you can use the four weights – 1, 3, 9 and 27 – to weigh everything from 1 gram to 40 grams.

MINUS 40 ...

degrees Celsius is exactly minus 40 degrees Fahrenheit. It's the only time the two scales correspond.

QUARANTENA

The word quarantine comes from the Italian (17th-century Venetian) word *quarantena* – meaning 'period of 40 days'.

F O R T Y

Forty is the only number in the English language where the letters are all in alphabetical order.

WD-40

Even the completely incompetent handyman or woman around the house knows about WD-40. Its makers proudly list over 2000 different things you can do with the wonder lubricant, from shining mailboxes to unsticking Lego blocks that are jammed together.

But you might not know that WD-40 was originally invented to protect the outer skin and fuel tanks of the Atlas ICBM Missile from rust and corrosion. Think about that next time you're unjamming some Lego blocks!

The name WD-40 comes from the abbreviation on the American chemist Norm Larsen's notes for 'Water Displacement 40th formula'. Well done, Norm, you persistent little chemist you.

功

MOZART

Many sources say he wrote 41 symphonies in his mere 35 years. Some scholars dispute he wrote them all, yet others suggest he wrote even more. One thing is for sure, no one calls Wolfgang a bludger.

WHAT THE PELL?

As we already discovered in chapter 29 when we met the Pell numbers, 41/29 = 1.41379 ... is a pretty good approximation for $\sqrt{2} = 1.41421$.

NEAT

While we are here, $41 = 4^2 + 5^2$ and look at this: $4^2 + 5^2 = 1^2 + 2^2 + 6^2$ and $4 + 5 = 1 + 2 + 6$.

Neat, hey?

GENIUS FROM BOREDOM

The Ulam spiral (or Ulam prime spiral) was famously discovered by Polish-American mathematician Stanislaw Ulam during a particularly boring maths lecture he was attending (what, a boring maths lecture? I know, I'm as surprised as you).

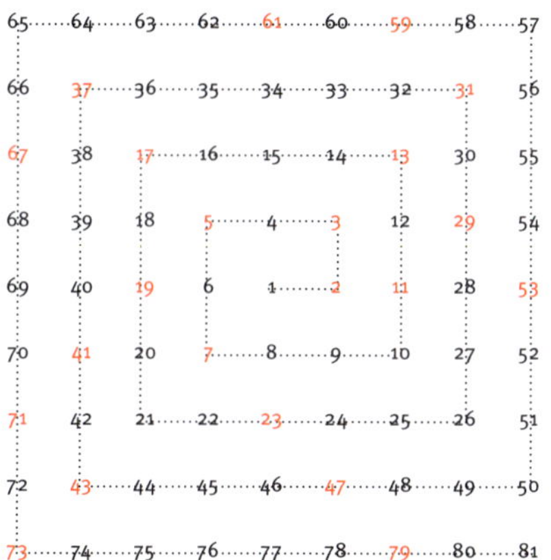

He placed the number 1 in the centre then began spiralling the counting numbers around it. But when he then looked for the primes they tended to occur along diagonals.

MY MATHEMATICAL MAN CRUSH

Leonhard Euler discovered literally thousands of things in his incredible lifetime. One of them was the formula $n^2 + n + 41$, which looks harmless but has the amazing property that, for whole numbers $n = 0$, $n = 1$, $n = 2$ up to $n = 39$, it gives a prime number as its value. What this means is:

$0^2 + 0 + 41 = 0 + 0 + 41 = 41$ is prime;
$1^2 + 1 + 41 = 1 + 1 + 41 = 43$ is prime;
$2^2 + 2 + 41 = 4 + 2 + 41 = 47$ is prime;
$3^2 + 3 + 41 = 9 + 3 + 41 = 53$ is prime;

all the way up to $39^2 + 39 + 41 = 1521 + 39 + 41 = 1601$

... which is also prime.

However, $40^2 + 40 + 41 = 1681$, which is 41×41. Good effort, Leonhard. And guess what – we can connect this equation of Euler's with an Ulam spiral that starts at 41 and spirals out:

				53
	45	44	43	52
	46	41	42	51
	47	48	49	50

Continue this until you get a 40×40 square with all of the primes along the diagonal. Then frame it and keep it on a wall as a monument to your hard work.

QUINTILLIONS OF POSSIBILITIES

'Enigma' was a machine used by the Germans in World War II to encode messages to their troops. When the operator typed a letter on the external keyboard an electric pulse passed through an extremely complex series of wheels, wires, sockets and rings to eventually produce a different letter to the one typed in.

But every time a letter was typed the machine recalibrated the wheels inside so the code would continue to change. If you typed 'G' 5 times in a row you'd get a different letter each time. To make matters worse, the next day, with the machine reset to a new starting code, those five 'Gs' in a row would come out differently again.

The only person who could read the code was someone else with an Enigma machine who had theirs set the same way as the person who typed the original message.

Combining the positions in which they set the wheels, rings and cables gave over 150 quintillion (150,000,000,000,000,000,000) different ways the Enigma machine could be set.

The 2014 movie, *The Imitation Game* introduced the great British mathematician Alan Turing to a new generation of cinema-goers. The fact this was nominated at the same Oscars as *The Theory of Everything* (about Stephen Hawking) pushed me into geek movie ecstasy.

Turing achieved so much breaking Nazis codes in World War II that Winston Churchill feted him as making the biggest contribution to the Allied victory. He also developed what is known as the 'Turing Test' which presaged the concept of artificial intelligence in computers and machines. Alan Turing died far, far too young at the age of 41.

While mathematical genius and German laziness played their parts in cracking the Enigma machine, also crucial were several occasions when the Allied forces managed to seize them after battles.

Though I should point out if you watch the movie *U-571* the Brits are mighty peed off that their soldiers actually did the legwork in that battle. America hadn't even entered the war at that stage!

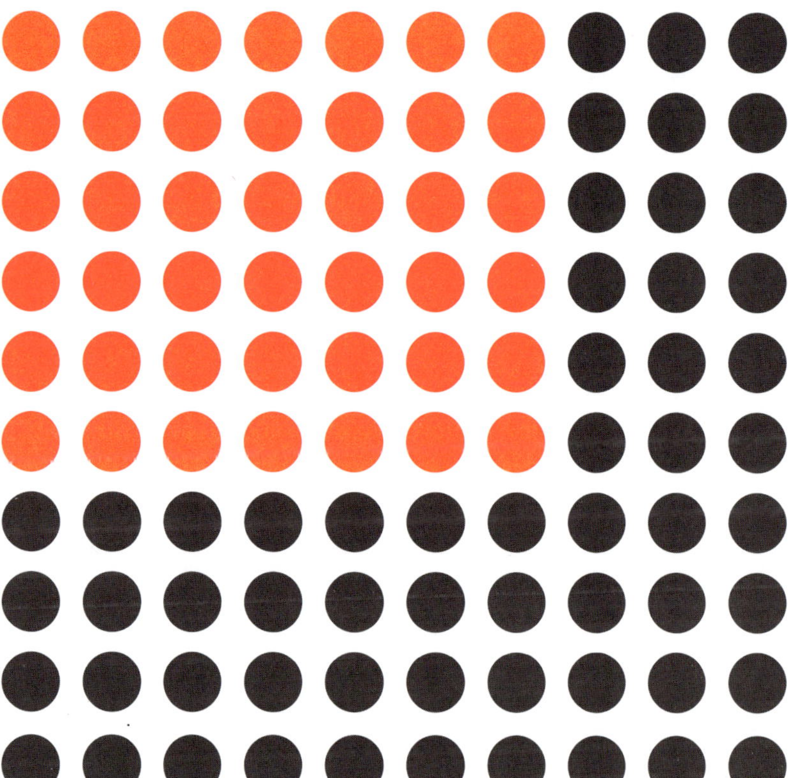

42

MORE ON THOSE COOL CATALANS

In chapter 14 we met the Catalan numbers 1, 2, 5, 14 ... They are given by a formula that looks scary the first time you see it, but isn't that bad:

$$C(n) = \frac{(2n)!}{n!\,(n+1)!}$$

We pictured the Catalan numbers as the number of ways you could break a polygon (square, pentagon, hexagon and so on) into internal triangles. Here are some more ways to picture the Catalan numbers.

They tell us the number of ways you can walk from the lower-left corner of a $n \times n$ grid to the upper-right corner without crossing above the diagonal and only going right or up:

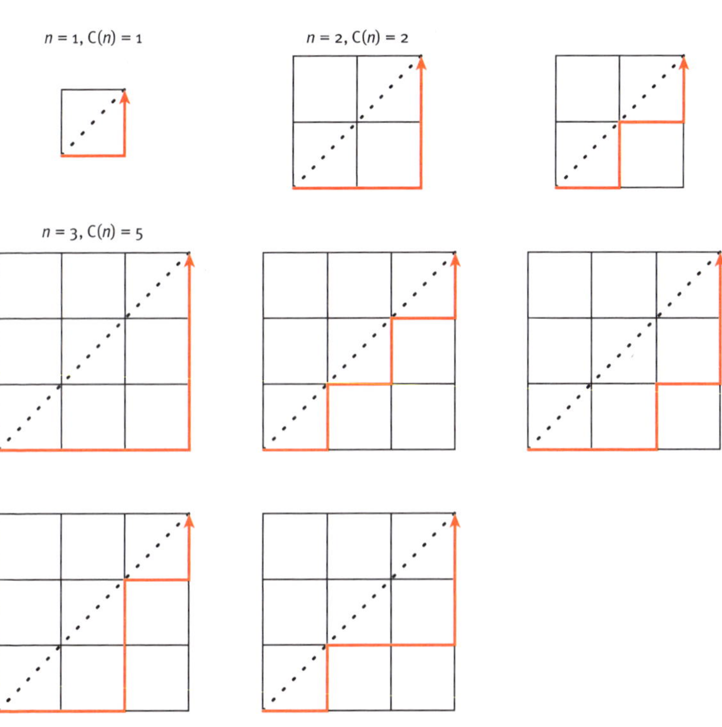

The Catalan numbers also tell us the number of ways you can assemble n Xs and n Ys into words, where reading from the left you never have more Ys than Xs (for example XXYYXY is fine but XYYXXY isn't because the first three terms XYY have more Ys than Xs).

$n = 1$, $C(n) = 1$	$n = 2$, $C(n) = 2$	$n = 3$, $C(n) = 5$
XY	XXYY	XXXYYY
	XYXY	XYXYXY
		XXYXYY
		XXYYXY
		XYXXYY

Both of these ways of counting give us the Catalan numbers.

It should be obvious that the steps and the words are linked. Consider X as a 'step right' and Y as a 'step up' and see how they match up.

Convince yourself that when $n = 4$ there are 14 ways of walking on the 4×4 grid and 14 acceptable words using 4 Xs and 4 Ys.

Then if you're feeling particularly brave, 42 is the fifth Catalan number; try and work out the 42 paths and 42 words. Ow!

THE MEANING OF LIFE

According to Deep Thought, the computer in Douglas Adams' *The Hitchhiker's Guide to the Galaxy*, the answer to Life, the Universe and Everything is 42. No-one really knows if this is true. But if it is, would the meaning of life become 46.2 under a 10% consumption tax?

HOURS OF FUN ...

We've already met partitions (for example chapters 4 and 9). Remember that $7 = 4 + 2 + 1$, $7 = 5 + 2$ and $7 = 3 + 1 + 1 + 1 + 1$ are all partitions of 7. Try to find all 42 ways of partitioning the number 10. To avoid repeating any, set your calculations out in a sensible order.

ANDREWS CUBES

On pages 65 and 66 of his 1917 book *Magic Squares and Cubes*, W. S. Andrews listed the four basic $3 \times 3 \times 3$ magic cubes.

In 3D they'd look like this:

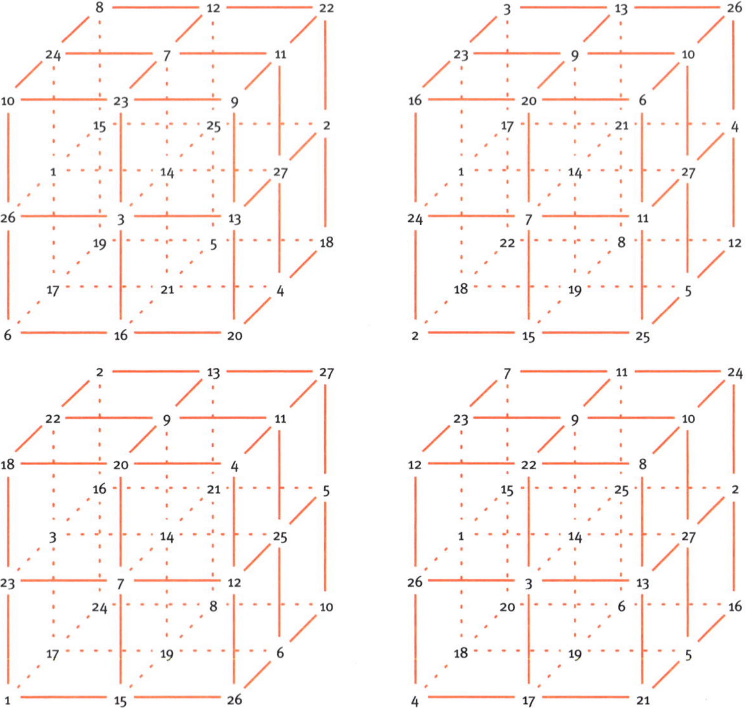

Every row and every column of every layer add up to 42 as do the rows and columns of the faces from front to back *and* the diagonals from a corner or edge through the middle to the opposite corner or edge.

The reason the magic constant is 42 is obvious when you think that the sum of all the numbers in the cube is $1 + 2 + 3 + \ldots + 26 + 27 = 378$ and there are three layers each with three lines, making 9 lines. So if each line is equal it must add up to $\frac{378}{9} = 42$.

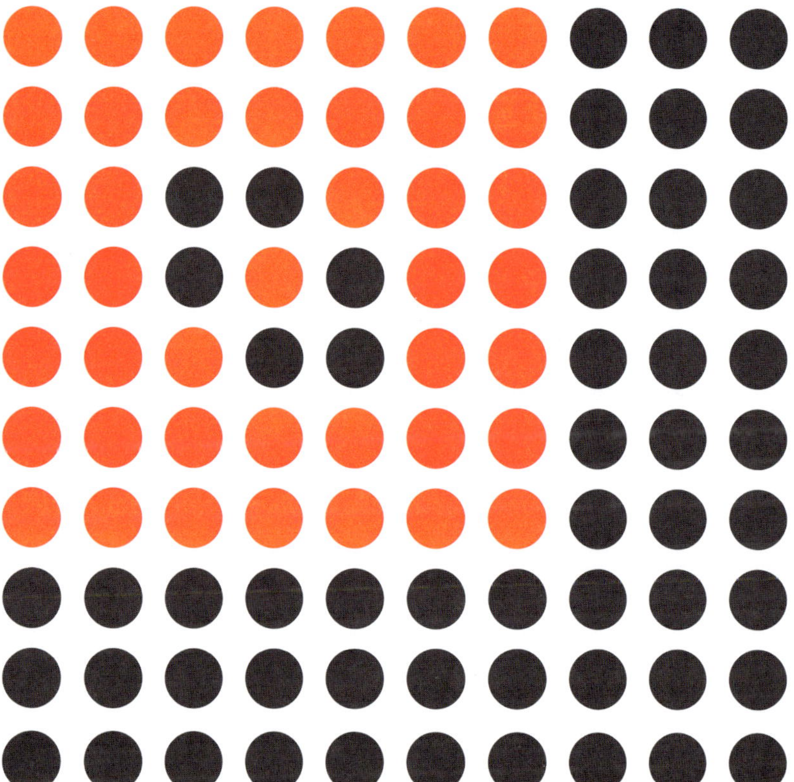

43

TWIN PRIMES

Forty-three is prime. In fact, 41 and 43 are consecutive odd numbers and are both primes. We call such pairs 'twin primes'. The twin primes are (3, 5), (5, 7), (11, 13), (17, 19) ...

In December 2011, we discovered the twin primes:

$$3{,}756{,}801{,}695{,}685 \times 2^{666{,}669} - 1 \text{ and } 3{,}756{,}801{,}695{,}685 \times 2^{666{,}669} + 1$$

... which are each 200,700 digits long!

We think that there are an infinite number of twin primes, but we haven't proven it yet. That said, in 2013, an until then virtually unknown middle-aged Chinese-American lecturer named Yitang Zhang, who had even had to make ends meet working in a Subway sandwich shop for a while, made a massive breakthrough which gets us much closer to confirming this long-suspected result.

QUIZ QUESTION: **Find all the twin primes between 1 and 100.** ANSWER AT THE BACK OF THE BOOK

Now look at the list of primes 3, 7, 11. It is a progression of 3 primes separated by a common difference of 4.

Similarly, look at 5, 11, 17, 23, 29. This is a progression of 5 primes separated by a common difference of 6.

Here is the really cool bit: in 2004, Aussie maths genius Terence Tao and colleague Ben Green proved that no matter how long you want your progression of primes to be, somewhere in the infinity of numbers, there is a progression of primes that long. Just think about it – you want a progression of 9 primes, it has to exist. Seventeen primes, it's out there. What about a progression of 7 trillion primes? While the Green-Tao theorem can't find the actual progression for you, it does prove that a progression of that length must exist!

In my humble opinion, that is beyond awesome.

Cuarenta y tres is Spanish for 43 and gives its name to the popular Licor 43, which is distilled from 43 different herbs and spices.

ON THE SUBJECT OF PRIMES

Forty-three is the smallest prime number expressible as the sum of 2, 3, 4 or 5 different primes. By that I mean:

$43 = 41 + 2$

$43 = 11 + 13 + 19$

$43 = 2 + 11 + 13 + 17$

$43 = 3 + 5 + 7 + 11 + 17$

RUBIK'S CUBE

The Rubik's Cube is probably the world's bestselling and – to most mathematicians – coolest toy ever. Most agree that the Cube was invented by Hungarian mathematician Ernö Rubik in 1974, probably to teach students at the Department of Interior design in Budapest about 3-dimensional objects. But as with all great inventions, other people have claimed to have had the idea first ... they just never got around to making it!

As we saw earlier, 43 is the number of million, million, million different ways to arrange a Rubik's Cube – give or take ... to be exact, it's 43,252,003,274,489,856,000 ways!

To try and explain how many possible arrangements this is, if we took standard Rubik's Cubes (just under 6 centimetres per side) in all possible arrangements and covered the Earth in them ... you'd cover the Earth 273 Cubes deep.

Or, if you laid them end to end, they'd stretch around 261 light years.

NON-MCNUGGET NUMBERS

So far, we've met some very important types of numbers – primes, Fibonacci numbers and so on. Now it's time to have a little fun.

When McDonald's Chicken McNuggets were first released they came in boxes of 6, 9 and 20. If you had a craving for McNuggets, you could order 6 nuggets, 12 (2 boxes of 6), 15 (a box of 6 and a box of 9), 18 (2 boxes of 9 or 3 boxes of 6), 20 (a box of 20) and so on.

But if you look at the options, boxes of 6, 9 and 20, you'll see it's not possible to order exactly 10 nuggets, for example.

And looking at all combinations of 6, 9 and 20 you'll see it's not possible to order exactly 43 nuggets.

Yet once we get to 44 (20 and 4×6), 45 (5×9), 46 ($2 \times 20 + 6$), 47 ($20 + 9 + 3 \times 6$), 48 (8×6) and 49 ($2 \times 20 + 9$), we notice this: we *can* buy 44, 45, 46, 47, 48 and 49 nuggets and we can buy more boxes of 6 nuggets.

So we can now order 50 (just use our order for 44 and add 6), 51 ($45 + 6$), 52 ($46 + 6$), 53 ($47 + 6$), 54 ($48 + 6$), 55 ($49 + 6$), 56 ($44 + 6 + 6$) and so on. Because from 44 to 49 we have 6 consecutive McNugget numbers and we have a box of 6 as an option, we now have all possible combinations from 44 up.

So, 43 is the highest non-McNugget number.

More generally, we say that 43 is the Frobenius number of {6, 9, 20}.

QUIZ QUESTION: Find all the non-McNugget numbers.

ANSWER AT THE BACK OF THE BOOK

QUIZ QUESTION: Find the Frobenius number of {4, 9} of {5, 8} of {6, 7} of {4, 7, 12} and, if you like a bit of a challenge, of {12, 16, 20, 27}. ANSWER AT THE BACK OF THE BOOK

カカ

THE INCLUSION-EXCLUSION PRINCIPLE

Four people, Ang, Blaise, Chantalle and Dragan, are all going to a party where they have to wear nametags. But they all thought it would be hilarious to wear the wrong nametag and confuse people. How many different ways could they all wear the wrong nametag?

This isn't too hard to work out by brute force. If we use the letters A, B, C and D the correct order of the nametags is (A, B, C, D).

If we gave out the nametags in the order (D, C, A, B) no-one would get the correct tag. We just want to work out all the sets like that with none of A, B, C or D in their correct spot.

Start by giving B to Ang and go from there.

(B, A, D, C) is one we're looking for, as is (B, D, A, C) and (B, C, D, A) and those are the only three ways Ang can wear Blaise's nametag and all the nametags be wrong.

I'll leave to you to prove that there are three ways Ang could wear Chantalle's nametag and three ways Ang could pretend to be Dragan.

So there are 9 ways they could all wear the wrong nametag.

What if Ellie joined the group and they all wanted to wear the wrong names?

For 5 people or more there are too many ways to just count them all out. The answer comes from a handy piece of mathematics called the 'inclusion-exclusion principle' which is used to compare the sizes of two or more sets of things when the sets may overlap in membership.

To use the inclusion-exclusion principle to solve this problem requires us to count the number of 'derangements' – a derangement is when you change the things in a set so that none of them is in their original position. To do this we use the formula:

$$\mathrm{d}(n) = n! \times \left(1 - \frac{1}{1!} + \frac{1}{2!} - \frac{1}{3!} + \frac{1}{4!} \dots + \frac{(-1)^n}{n!}\right)$$

Now, don't freak out! I've put this in the book precisely to show you that really weird mathematical notation – like the sort of stuff they write on blackboards in cartoons when the crazy professor is explaining how she will destroy the world – is pretty harmless stuff. I'll walk you through this.

We know what the ! sign means after a number. It's just the factorial sign, where $4! = 4 \times 3 \times 2 \times 1$.

So what we do here is set $n = 5$ (the 5 people at the party) and then add up all the fractions inside the brackets until we get to $n = 5$.

So for all of its bluff and bluster ...

$$d(5) = 5!\,(1 - \frac{1}{1!} + \frac{1}{2!} - \frac{1}{3!} + \frac{1}{4!} - \frac{1}{5!})$$
$$= 120\,(1 - 1 + \frac{1}{2} - \frac{1}{6} + \frac{1}{24} - \frac{1}{120})$$

we write all the fractions in the big bracket over 120 and get:

$$d(5) = 120\,(\frac{120}{120} - \frac{120}{120} + \frac{60}{120} - \frac{20}{120} + \frac{5}{120} - \frac{1}{120})$$
$$= 120 \times (\frac{44}{120})$$
$$= 44$$

So there are 44 ways 5 people can all wear a wrong nametag.

Well done!

The principle was discovered by the brilliant French mathematician Abraham de Moivre who also assisted Euler in giving us the identity:

$$e^{i\pi} + 1 = 0$$

... which is irrelevant here, but so beautiful it needs no justification.

Now, confront your demons and try this. Using the formula for $d(n)$, show that for 6 people we get 265, for 7 people 1854 and for just 8 mischievous dinner party guests there are 14,833 derangements!

Though we have in our noses an impressive 5 million olfactory cells with which to smell, sheepdogs have 220 million, so they can smell 44 times better than we can. However, only some of us smell 44 times nicer than the average sheep dog.

MASSIVE PRIME NUMBER

Mathematicians love prime numbers and have known for thousands of years that the primes never stop: there are an infinite number of them. My TED Talk on the subject is at http://bit. ly/1wbWfXs.

Back in chapter 31 we met the gigantic prime number $2^{57,885,161} - 1$ which has almost 17.5 million digits and as I explained earlier if typed out would run the length of the 7 Harry Potter books and about half again!

But before the age of computers, people were making amazing calculations with pen and paper or just basic calculators.

In 1951, A. Ferrier announced that, with only a desktop calculator, he had proven $(2^{148} + 1) \div 17$ to be prime. The 44-digit number:

20,988,936,657,440,586,486,151,264,256,610,222,593,863,921

... was the last pre-computer age 'biggest prime'.

In a truly unfair case of timing, barely a month later, the number $180 \times (2^{127} - 1)^2 + 1$ ... which is 79 digits long, was shown to be prime and the computer age of prime detection was off and running.

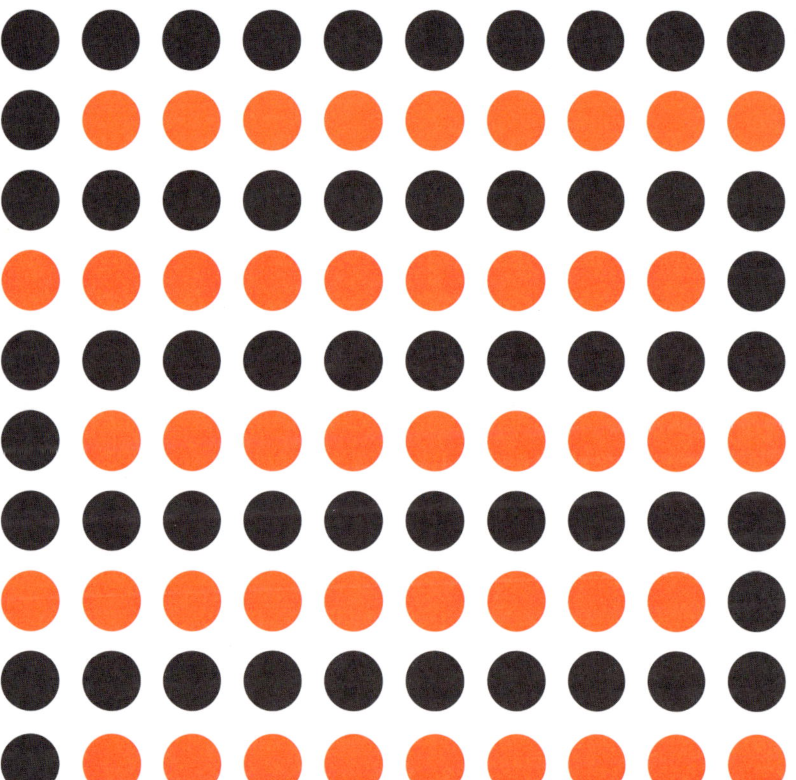

45

LUCKY BALLS?

Forty-five is the number of balls used in many 'lotto' draws around the world so let's calculate the odds of ever winning 'the big one'.

We'll start with a straight draw and move onto 'power balls' and the like.

Assume there were only balls 1, 2, 3 and 4 in the lotto draw and you had to get 2 from 2 correct to win the jackpot. Now, I know that doesn't seem too hard – that's why they use 45 balls not 4 – but if you can follow the maths here it's easy to ramp it up for the bigger draw.

There are 12 ways the 2 balls could come out:

(1 2) (1 3) (1 4) (2 1) (2 3) (2 4) (3 1) (3 2) (3 4) (4 1) (4 2) and (4 3)

But if you pick (2 3) and the balls come out (3 2) you still win, so in fact every 2 ball choice can come out 2 different ways and still give the same result. Thus our 12 possible results collapse to just $12 \div 2 = 6$:

(1 2) (1 3) (1 4) (2 3) (2 4) (3 4)

Choosing 6 balls from 45 makes the numbers bigger, but essentially there are:

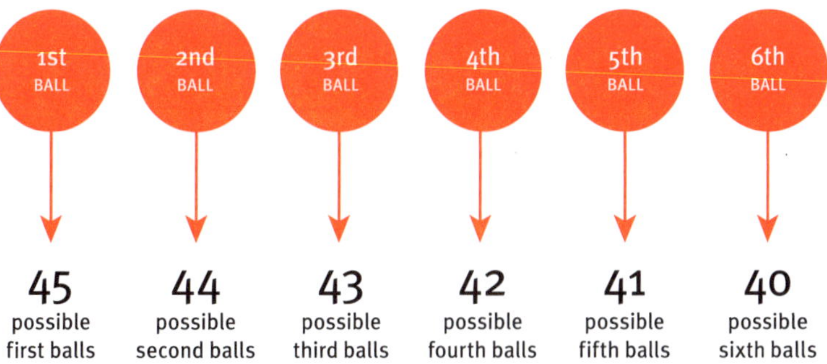

... $45 \times 44 \times 43 \times 42 \times 41 \times 40$ ways the 6 balls could come out.

But any choice of 6 balls that you make could come out $6 \times 5 \times 4 \times 3 \times 2 \times 1$ different ways, so the number of possible results collapses to:

$$\frac{45 \times 44 \times 43 \times 42 \times 41 \times 40}{6 \times 5 \times 4 \times 3 \times 2 \times 1}$$

... that's 8,145,060 possible choices. Thus the odds of winning that lotto are 1 in 8,145,060.

You should also be aware that something as innocent sounding as changing it to 'get 5 from 5 correct and nominate the 6th number which will be the powerball' alters those odds to a whopping 48,870,360 to 1.

Good luck, you'll need it.

CONSECUTIVE DIGITS

Forty-five can be written as the sum of consecutive digits 5 different ways. For example $45 = 22 + 23$ and $45 = 5 + 6 + 7 + 8 + 9 + 10$.

QUIZ QUESTION: Find the other 3 ways that 45 can be written as the sum of consecutive digits. ANSWER AT THE BACK OF THE BOOK

A laboratory in Minnesota in the US has an 'anechoic chamber', a room insulated against all noise which is said to be the quietest room in the world. Interestingly, the longest anyone has managed to stay in the room is 45 minutes.

The body of the average adult human contains 45 litres of water; as much carbon as a 12-kilogram bag of coke (the black stuff); 2200 matchheads worth of phosphorus; enough iron to make a 25-millimetre nail; and enough lime to whitewash a small shed. Note, none of these things are easy, painless or legal to extract from your average adult human.

KAPREKAR NUMBERS

Forty-five is the third Kaprekar number. When you square an *n*-digit Kaprekar number and add the *n* digit number on the right to the *n* or $n - 1$ digit number on the left, you get the original number, so:

$45^2 = 2025$ and $20 + 25 = 45$

If the square of the number has an odd number of digits, split it so as to make the number on the right the longer number. So 2223 is Kaprekar because $2223^2 = 4,941,729$ and $494 + 1729 = 2223$.

QUIZ QUESTION: Which of the following are Kaprekar numbers?
51, 55, 59, 91, 99, 103, 295, 297 ANSWER AT THE BACK OF THE BOOK

Kaprekar numbers can also be defined for cubes, and higher powers.

For example:

$45^3 = 91,125$ and splitting 91,125 into 3 parts we get $9 + 11 + 25 = 45$ so 45 is a Kaprekar number for the third power.

And guess what ... no ... go on ... guess ... okay, I'll tell you:

$45^4 = 4,100,625$ and $45 = 4 + 10 + 06 + 25$, so 45 is also a Kaprekar number for the fourth power! It is also the only known number (apart from the trivial example of 1) that is in the Kaprekar sequences for squares, cubes and fourth powers.

QUIZ QUESTION: Which of the Kaprekar numbers from the first Quiz Question is also a Kaprekar number for the fourth power? ANSWER AT THE BACK OF THE BOOK

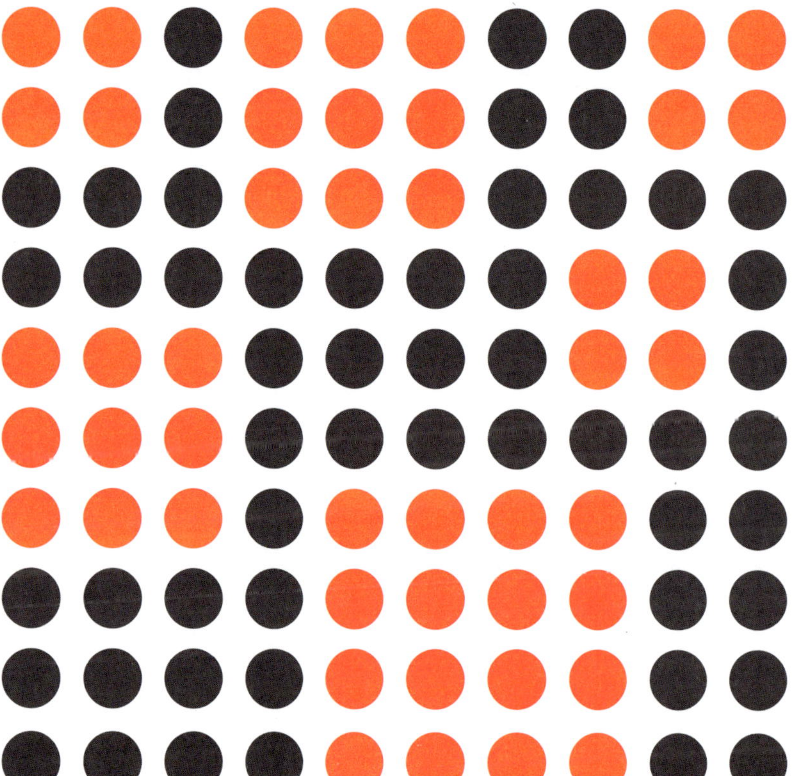

46

SUPER STAR

This amazing hexagram contains the numbers 1 to 19 and every 5-number diagonal adds to 46. Martin Gardner showed me this in his gorgeous book *Are Universes Thicker Than Blackberries?*

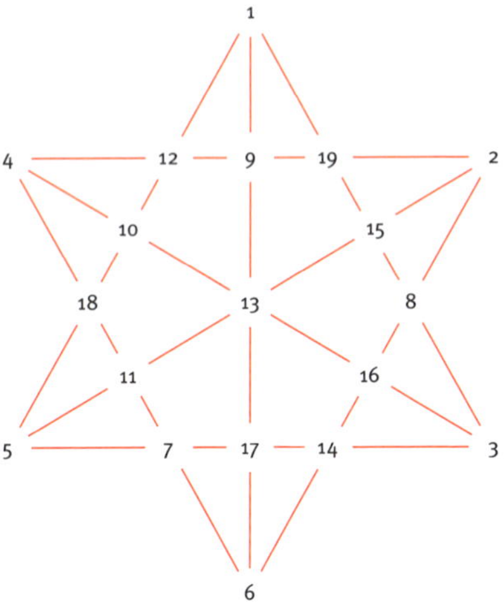

BARDY BIBLE

I like a good conspiracy theory as much as the next guy – Moon landing ... shot on a Hollywood back lot – weather balloons ... that's what you'd expect the government to say, after all, they *are* aliens too.

But this one I particularly love.

Shakespeare clearly wrote the Bible. Or at least parts of it.

The King James Bible was finished in 1611. In 1610, Shakespeare was 46 years old. Go to Psalm 46. Count 46 words in ... you find

'the mountains will *shake* with the swelling thereof'

and count back 46 words from the end ... you find

'and cutteth the *spear* in sunder'.

Stop it, dude – you're freaking me out. Shakespeare is 46 in 1610 and 46 words from each end, of Psalm 46, says 'Shake-spear'.

To many it is clear proof that King James enlisted the Bard to help translate his version of the Bible. To me, it is what we mathematicians call ... a coincidence. I'll let you be the judge.

HOW LONG IS A YEAR?

You probably answered '365 days' or '365 or 366, depending on whether it's a leap year'. Perhaps you even know that the solar year (the time taken for the Earth to fully orbit the Sun) is roughly 365 days, 5 hours and 49 minutes – and I stress that's only roughly. But it hasn't always been that accurate.

Around 700 BC, the Romans were using a 10-month, 355-day calendar. But it was obvious to the farming community that this was about 10 days short of the mark, because the seasons soon fell out of sync with the weather. To overcome this, the Roman priests and aristocrats tried various techniques based around inserting extra months now and then to top up the days that had been lost.

The system didn't work too well due to a combination of factors: people were often confused about when the extra months were due, it was often politically expedient to muck around with the process and change the dates that ceremonies and feasts were due, and the big cheese himself often had more pressing matters to address – I mean, if you had to choose between spending a romantic weekend with Cleopatra or deciding whether February should be a few days longer, which way would you go?

By 46 BC, the situation was way out of hand, and the calendar had slipped from the solar year by almost 2 months. Caesar rectified the situation in one fell swoop – he added the extra 23 days already due for February and another 2 months, of 33 and 34 days, between November and December. As a result, 46 BC, which Caesar called '*ultimus annus confusionis*' (the last year of confusion) but most Romans just called the Year of Confusion, lasted a whopping 445 days.

There are 46 species of birds in Antarctica, all of which, I'm presuming, are pretty hardy critters.

IT'S STILL HAILING

We met the hailstone problem in chapter 27. You'll recall (well, to be honest I don't expect you to, you might have to go back and check but that's fine) that you take any number and halve it if it's even and triple it and add 1 if it's odd. No matter where you start you'll eventually end up with the number 1.

But what can vary wildly is the number of steps it takes.

For the number 46 we get:

46 goes to 23, 70, 35, 106, 53, 160, 80, 40, 20, 10, 5, 16, 8, 4, 2, 1

... just 16 steps to take us to 1.

Convince yourself that 44 and 45 both reduce to 1 after exactly 16 steps.

But just when you think there might be some sort of pattern forming here, along comes 47 which bounces around for ages at one stage getting as high as 9232 before finally reducing to 1 after 104 steps.

47

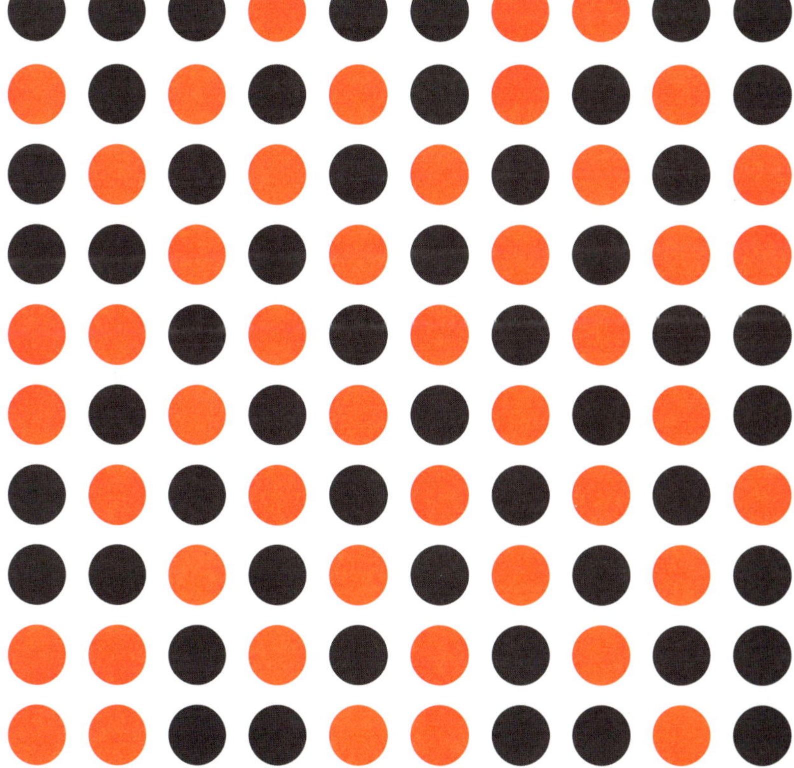

CUNNINGHAM CHAIN

Take the number 2, double it and add 1. What do you get? That's right, 5.

Take 5, double it and add 1. What do you get? 11? Spot on.

Keep doing this and what do you notice about the first 5 numbers you get?

2, 5, 11, 23, 47 ... they are all prime numbers.

So we call the chain 2, 5, 11, 23, 47 a 'Cunningham chain' of length 5.

Notice that if we double 47 and add 1 we get 95, which is not prime.

QUIZ QUESTION: There is a pretty impressive Cunningham chain of length 6, which begins with a number in the 80s. Find the chain.

ANSWER AT THE BACK OF THE BOOK

ULAM NUMBER

In chapter 41 we met the fascinating Polish-American Stanislaw Ulam. Among other things, he worked on America's Manhattan Project, the World War II campaign to build an atomic bomb at Los Alamos in New Mexico.

Forty-seven is an Ulam number, named after ... you'll never guess who. We get the Ulam numbers by writing 1, 2, 3 and then any other numbers that can be written *only once* as the sum of 2 other numbers on the list. So $1 + 3 = 4$ and we put 4 on the list, but once we have 1, 2 3, 4 we don't include 5 because $5 = 1 + 4$ *and* $5 = 2 + 3$. The list of Ulam numbers begins 1, 2, 3, 4, 6, 8, 11, 13, 16, 18 ...

QUIZ QUESTION: Find the Ulam numbers up to 100. ANSWER AT THE BACK OF THE BOOK

47 STRINGS

A violin has 4 strings and a standard guitar has 6 (no offence to bass guitarists or those who like to double string for 12 strings). Pianos have in the range of 230 strings – it depends on the make and model as to which notes are provided by 1 2 or 3 strings – most keys have 3 strings attached.

But you don't need hundreds of strings to make beautiful music. The Chinese erhu has only two strings, and the oud, the beautiful Turkish instrument played by my friend (and one of the best oudists in the world) Joseph Tawadros, has 11 strings – grouped into 5 pairs and a single lowest note.

How many strings does a harp have? The answer to that seemingly simple question is actually quite complicated. Essentially, it depends what culture and what sort of harp you're talking about.

An Afghani Kafir harp has just 4 or 5 strings, various Tamil yaal harps have anything from 7 to 21, and Paraguay's national instrument, that's right the harp, has 32, 36, 38, 40, 42 or 46 strings.

The Pedal harp or Concert harp seen in western chamber ensembles has 46 strings ... (except of course for the ones that have 47).

KEITH NUMBER

We've already encountered the Fibonacci numbers (see chapters 3, 8, 13 and 19). Well, American engineer Mike Keith gives his name to numbers with a property based on a Fibonacci-like process. Huh? Trust me here.

To see if an *n*-digit number is a Keith number, write out the sequence that starts with the n-digits of the number. Then to get each new each term, add the previous *n* terms.

For example, 197 is a 3-digit number so we form the sequence:

1, 9, 7, 17 (1 + 9 + 7), 33 (9 + 7 + 17), 57 (7 + 17 + 33)z, 107, 197 ...

Because 197 appears in its own sequence, 197 is a Keith number.

They are pretty rare but if you'd like why don't you ...

QUIZ QUESTION: Prove 47 is a Keith number and find all other Keith numbers less than 100. HINT: Two are in the teens, one in the late 20s, one in the 60s and one in the 70s. ANSWER AT THE BACK OF THE BOOK

47 TRIANGLES IN A TRIANGLE

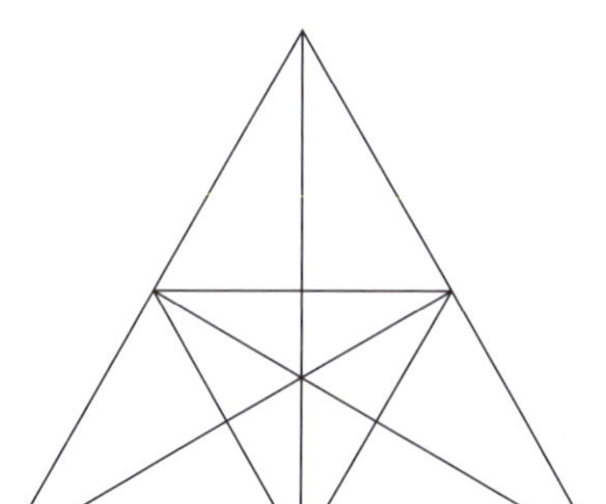

There are 47 smaller triangles contained in this larger one. But you might want to take my word on that. You can try to find them but it's a few hours of your life you won't be getting back!

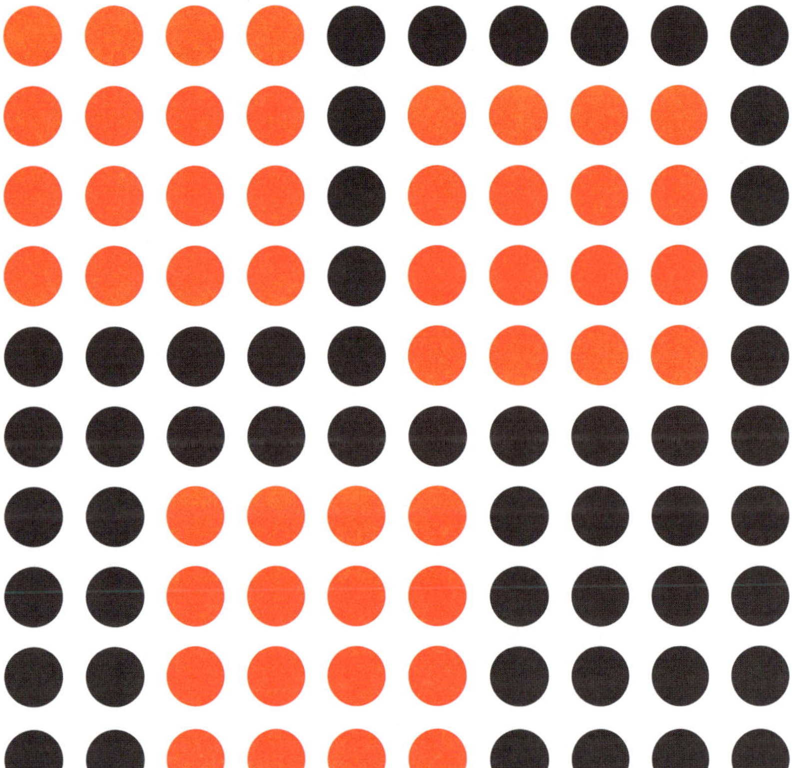

48

A GREAT RHOMBICUBOCTAHEDRON

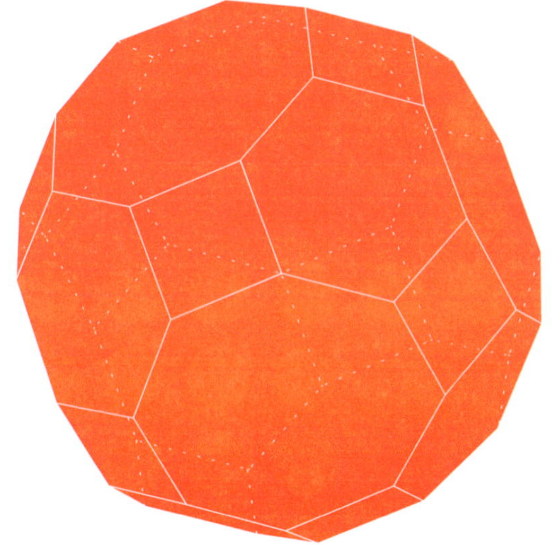

… has 26 faces, 48 vertices and 72 edges. Check that Euler's formula $V + F - E = 2$ applies to this Archimedean solid.

And it's also pretty impressive when you see how it would look flattened out:

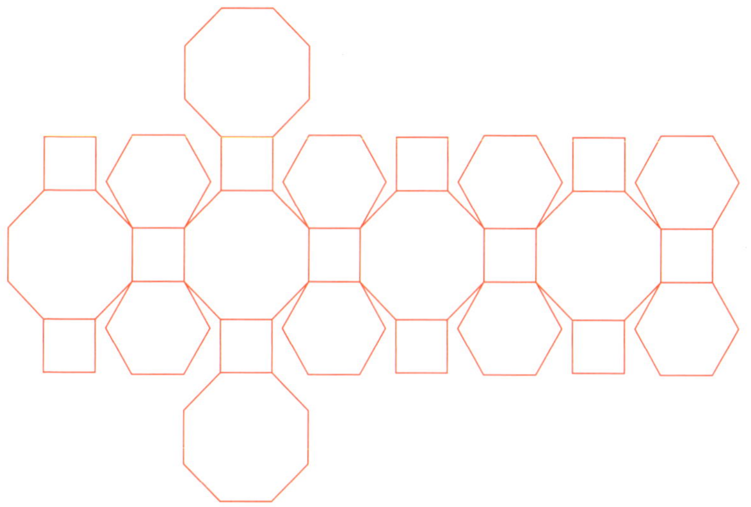

48 ?!*@$! HOURS

In the 1982 cinema (not quite) classic *48 Hours*, Eddie Murphy drops the f-bomb 48 times in its various incarnations. I'm surprised the movie wasn't called *48 Motherf****n Hours*.

SPEAKING OF 48 HOURS

one of Charlie Sheen's more memorable roles was a drug addict at the police station in *Ferris Bueller's Day Off*. To prepare for the gig, he didn't sleep for two days. Charlie Sheen has stayed awake for more than 48 consecutive hours many times in his life, but this is the only known instance of him doing it for artistic reasons.

POTATOES

have 48 chromosomes, 2 more than most human beings. Which just goes to show you that chromosomes are a start, but they're not everything.

Forty eight is also abundant (see chapter 12), highly composite (see chapter 36) and Harshad (which you'll see later in chapter 84 – I can't wait, can you?)

HAPPILY EVER AFTER

Forty-eight and 75 are 'betrothed'. This happy couple have a relationship because the factors of 48 are 1, 2, 3, 4, 6, 8, 12, 16, 24 and 48. The sum of the factors of 48, except for 1 and itself, is $2 + 3 + 4 + 6 + 8 + 12 + 16 + 24 = 75$. The sum of the factors of 75, except 1 and itself is, you guess it, 48. How cute!

QUIZ QUESTION: Which of the following pairs are also betrothed? 84 and 88; 140 and 195; 108 and 171; 1575 and 1648. ANSWER AT THE BACK OF THE BOOK

YOU SAY AWARI, I SAY OWARE

Oware, sometimes called Awari, is a beautifully simple but highly addictive board game that is very popular in West Africa and the Caribbean. I was first taught it by a friend from Ghana where it is the national game. It's a great way to learn maths and when it's played publicly, spectators are encouraged to give advice to both players.

Like a lot of great games it is simple to learn but extremely hard to master.

It starts with 48 seeds split 24 each between the 2 players, 4 seeds in each of their 6 'houses'. To move you simply pick up all the seeds in one of your houses and 'plant' them one at a time around the board counterclockwise.

Each time your last seed lands in an opponent's house that has only 1 or 2 seeds, you get all the seeds in that house. If the previous-to-last house also has only 1 or 2 seeds in it as you come round, you empty that house as well ... and so on.

The first person to capture 25 seeds wins.

In 2002, mathematicians using computers showed that if two people play the perfect game of Oware it will always end in a draw. But if the person going first empties anything other than the house closest to their opponent (house 6) a 'perfect' opponent will beat them.

Thankfully, humans aren't perfect, so Oware is still tremendous fun.

Trust me, if you get a chance to play this you should – it rocks.

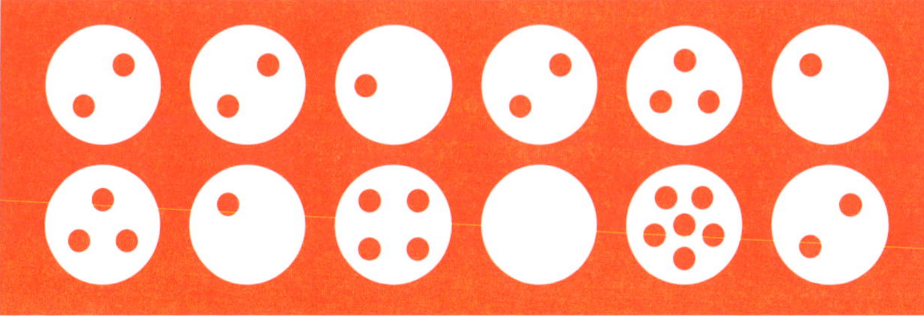

67

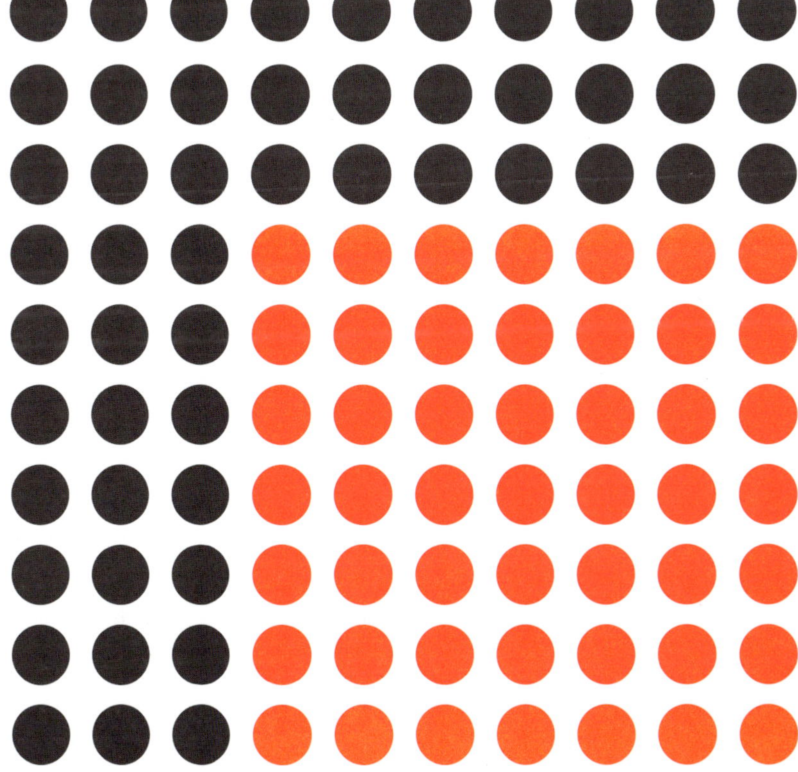

KNOTS

Knot theory sits in a part of mathematics called 'topology'. One thing topology looks at is if shapes can be bent and stretched into other shapes without tearing them or gluing them back together. Huh?

To a topologist, a watermelon and a die are essentially the same. They each exist in 3 dimensions, have one surface an ant could crawl all the way over, and no hole through them.

But to a topologist, an orange and a doughnut are essentially different. The hole through the centre of the doughnut means that while you could bend and stretch a sphere into a cube, you could never turn a sphere into a doughnut without ripping the surface and tearing a hole through it.

Knot theory looks at mathematical objects called 'knots'. Think of them as shapes you could get by taking a piece of string, wrapping it around itself and perhaps through loops, and joining the ends together when you're done. Mathematicians like to look at various knots that might not look the same and ask whether they are essentially different or really just the same knot.

The most simple knot is strictly the 'unknot' – which looks like the one in Diagram 1.

Diagram 1

Diagram 2

Diagram 3

Diagram 4

The first interesting (or non-trivial) knot is the 'trefoil' knot. It has 3 crossings (Diagram 2).

Diagram 5

Two things to notice here are first that there are no knots with only one or two crossings. That's because knots like the ones in Diagram 3 can be twisted into an unknot without cutting and rejoining them anywhere. So, as far as knot theorists are concerned – hey, don't laugh, some of my best friends are knot theorists – knots that look to have 1 or 2 crossings are just unknots.

Diagram 6

The second interesting thing is that there is only one trefoil knot. That is because this knot (Diagram 4) is a mirror image of the knot in Diagram 5.

As the number of crossings increase, things get interesting. There is only one type of knot with 4 crossings, shown here in Diagram 6, but for 5 crossings, these 2 knots are different (Diagram 7). You can't slide or twist them to get each other.

Diagram 7

After that things go wild fast. There are 3 different knots with 6 crossings and 7 with 7 crossings. There are 21 knots with 8 crossings and 49 with 9 crossings (an example of which is Diagram 8).

By the time we count knots with 10 crossings, there are 165 of them which, thankfully, takes us well beyond the purview of this book.

Diagram 8

On 25 September 2000, in front of 112,524 people, Cathy Freeman ran her guts out for 49.11 seconds and into Australia's heart when she won the Womens' 400 metres final at the Sydney Olympics.

HIP TO BE 49

Forty-nine is a perfect square: $7 \times 7 = 49$. In fact, both its digits, 4 and 9, are square, as is their product: $4 \times 9 = 36$.

On the topic of 49 and squares, watch, learn and then impress your friends:

$7 \times 7 = 49$

$67 \times 67 = 4489$

$667 \times 667 = 444{,}889$

$6667 \times 6667 = 44{,}448{,}889$. Can you see the pattern?

QUIZ QUESTION: What's $666{,}667^2$? ANSWER AT THE BACK OF THE BOOK

LIBYA

In September 2011, the United Nations recognised the name 'Libya' for the African nation that had until then most recently been known as 'Great Socialist People's Libyan Arab Jamahiriya'. If you think that was a mouthful, in Arabic it was

al-Jamāhīriyyah al-'Arabiyyah al-Lībiyyah ash-Sha'biyyah
al-Ishtirākiyyah al-'Uẓmá

... at 68 letters the world's longest official country name. The title now passes to what most people call Algeria but is officially Al Jumhuriyah al Jaza'iriyah ad Dimuqratiyah ash Sha'biyah with 49 letters just tipping out The United Kingdom of Great Britain and Northern Ireland.

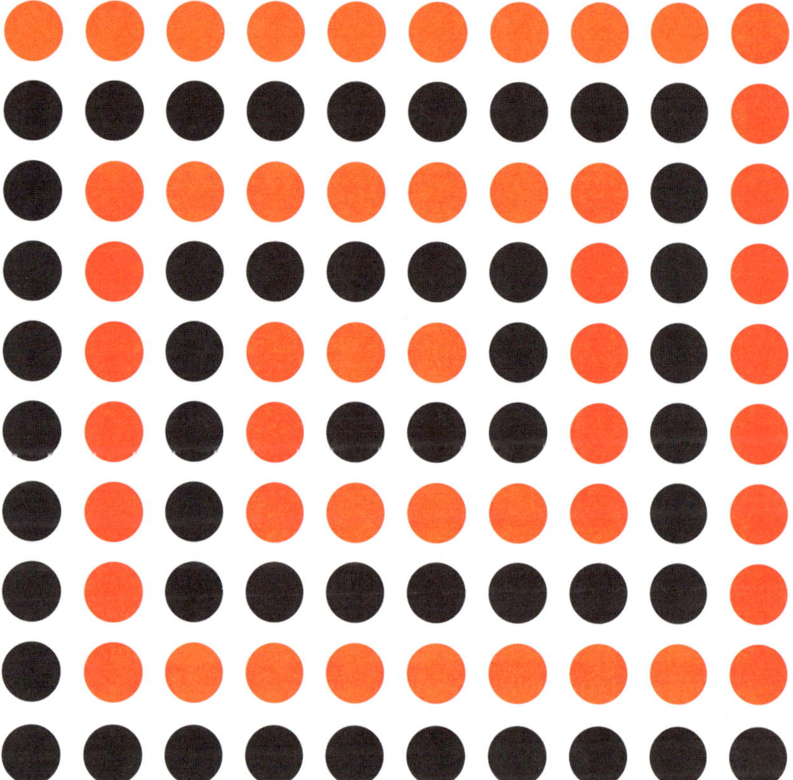

50

Fifty is the smallest number that is also the sum of 2 squares in 2 different ways. So $5^2 + 5^2 = 50$ and $7^2 + 1^2 = 50$. (Note, I'm ignoring the case of $25 = 3^2 + 4^2 = 0^2 + 5^2$ here.)

We can calculate this easily. Over the page, we'll see what it means 'geometrically'.

QUIZ QUESTION: What is the next number that can be written as the sum of 2 squares, 2 different ways? HINT: It's less than 100.

ANSWER AT THE BACK OF THE BOOK

TA DA!

There's a famous piece of 'magic' that regularly gets around the internet and completely blows people's minds. I know this 'cause I regularly get emails saying, 'Hey, I know you've done some mathematics ... can you PLEASE explain how this works. It's FREAKING ME OUT!'

Here is the 2015 version of the 'magic':

Think of a number, then multiply it by 2.

Now add 5.

Multiply that by 50.

If you've already had a birthday this year, add 1765, if you haven't had a birthday yet, add 1764.

Now, take away the year you were born.

The number you have left is made up of the number you thought of ... and your age!

Try it yourself right now ... and prepared to be blown away!

It's not that hard to see what is happening here.

Let's take any number X to start and let's say you've had your birthday already in 2015.

Multiply it by 2: this gives us $2X$

Add 5: this gives us $2X + 5$

Multiply by 50: we arrive at $50 \times (2X + 5) = 100X + 250$

Add 1765: we get $100X + 250 + 1765 = 100X + 2015$

Subtract the year in which you were born ...

$100X + 2015 -$ (the year you were born)

But if you're doing this in 2015, then '2015 minus the year you were born' just gives ... your age!

If you are 29 and you chose 8 to start, we have $100 \times 8 + 29 = 829$

If you are 16 and chose 39, we have $100 \times 39 + 16 = 3916$

It always comes out like this – the number you chose and your age. It's not magic; it's just that the numbers and the steps of the puzzle are carefully chosen to cancel each other out and leave you with your number and your age.

As a general rule, if someone was going to summon up 'magic', it would be a bit smoother than, 'If you've already had a birthday this year subtract 1765, otherwise subtract 1764 ...'. That should be a bit of a giveaway that the fix is in.

SEEING THROUGH GEOMETRY

Here's a neat piece of geometry to show $50 = 5^2 + 5^2 = 7^2 + 1^2$.

Let's fill a circle up with squares:

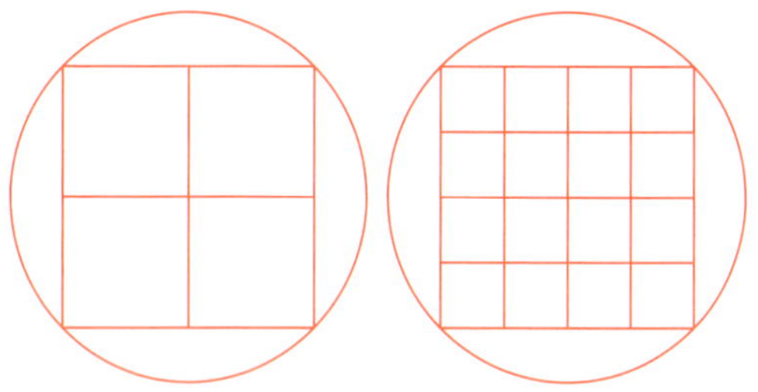

In both of these cases I couldn't fit another of the squares inside the circle. But when I place a 5×5 square inside a circle I can actually squeeze in 4 more squares:

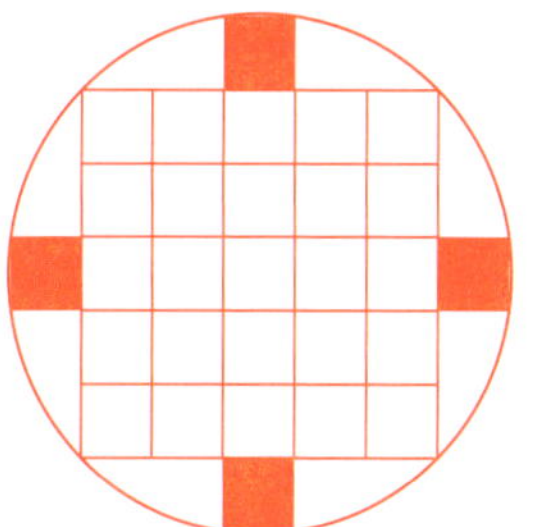

And because we know these two diameters are equal, our old friend Pythagoras tells us that $5^2 + 5^2 = 7^2 + 1^2 = 50$.

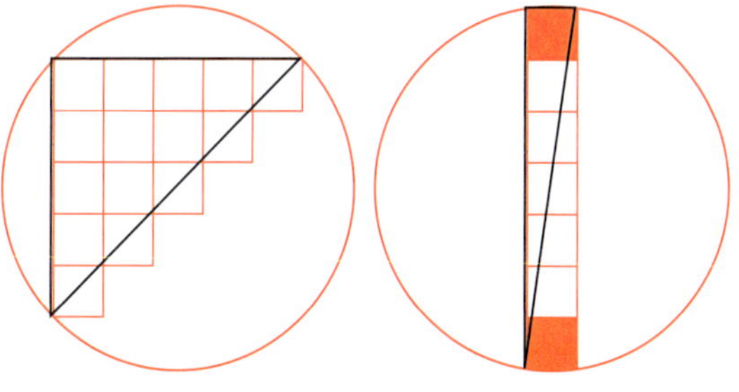

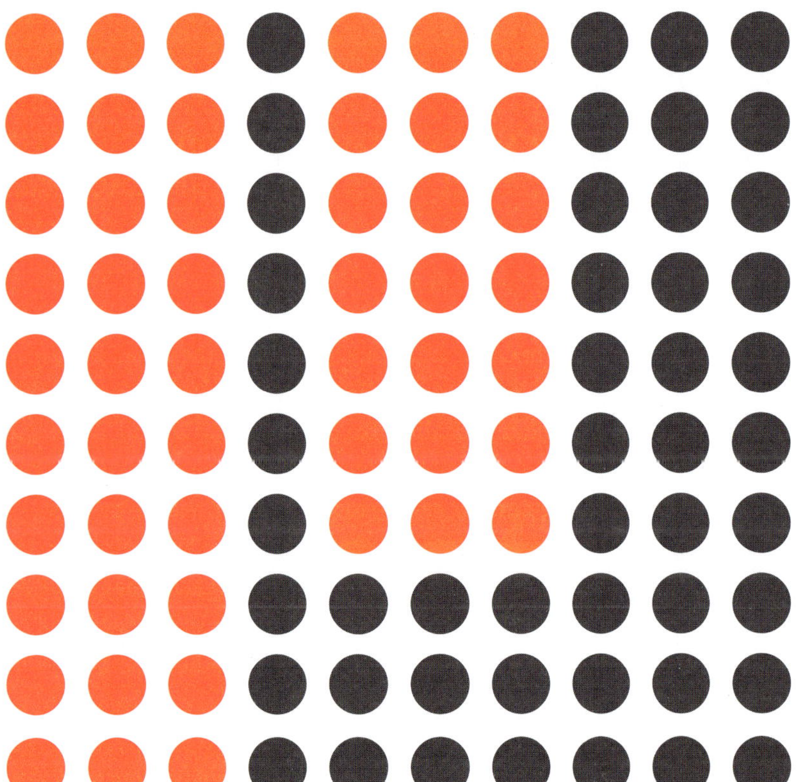

51

YOU CAN HANDLE THE PROOF!

The number 51 is divisible by 3. In fact, $51 = 3 \times 17$.

That itself isn't so amazing, but in my experience when you get someone to list the prime numbers, 51 is often one they get wrong. It's understandable because it doesn't feature on the times tables that you learn at school which usually finish at 12. But it's actually very easy to tell if a number is divisible by 3.

We've already mentioned how to do this in chapter 3.

If the digits of the number add up to something divisible by 3 the number itself was divisible by 3.

Back in chapter 3 we were just getting to know each other and I didn't want to scare you away, so I just let that fact sit by itself. But now that you're up to chapter 51 I think you can handle the proof.

To see why this is true you need to remember why we write our numbers the way we do.

For example, 762 is written the way it is because it is 7 lots of 100 and 6 lots of 10 and 2 lots of 1; or $762 = 100 \times 7 + 10 \times 6 + 2$. So the 3-digit number $ABC = 100A + 10B + C$.

It should be obvious that $9 \times B$ is divisible by 3 because 9 is divisible by 3 and that $99 \times A$ is divisible by 3 because 99 is divisible by 3.

And it should also be obvious that if we add together a group of things that are all divisible by 3 the result will also be divisible by 3.

Let's assume that $A + B + C$ is divisible by 3.

So $(A + B + C) + 9 \times B + 99 \times A$ will be divisible by 3 because it is just three things added together that were all divisible by 3.

But $(A + B + C) + 9 \times B + 99 \times A = 100A + 10B + C = ABC$.

So, if $A + B + C$ is divisible by 3, so is ABC. There you go, you could handle it, after all.

The growth of something that doubles (or triples, quadruples ...) each time is called 'exponential growth' and it is so fast it often stuns people.

If you could fold a piece of paper 51 times the rectangle would become *very* small ... but would be 2^{51} or over 2,250,000,000,000,000 times its original thickness and would stretch to beyond the Sun.

Chiron, or as it's officially known, 'Minor Planet 2060 Chiron', was first seen in 1895 and orbits the Sun every 51 years out between Saturn and Uranus. In 1979, Chiron was added to the fairly recently created category of rocks in space called 'centaurs'.

THE LITTLE MASTER

One of the greatest batsmen in the history of Test cricket is India's Sachin Ramesh Tendulkar, also known as 'The Little Master'.

Sachin made his Test cricket debut aged only 16 and in his 200 Test matches, wracked up an incredible 51 Test centuries – from his 119 not out against England in August of 1990, through to his 146 against South Africa in January 2011.

On the way, he clobbered Australia in a sublime 241 not out off 436 balls, which he compiled in 613 minutes ... and which I was lucky enough to watch at the Sydney Cricket Ground in 2004.

STOP, BUBBLE TIME.

The biggest bubblegum bubble blown had a diameter of a touch under 51 centimetres. The record was set by Chad Fell at the Double Springs High School, Alabama, US on 24 April 2004. Nice one, Chad.

Now, before any Chad Fell fans try and get hold of me, I should point out that this page's background shot is *not* a photo of said Mr Fell and his world record bubble.

Don Adams as Maxwell Smart on his famous shoe phone. Source: Wikipedia

A running gag in the 1960s spy spoof *Get Smart* was the hidden phone. There were 51 hidden phones over 138 episodes. In addition to his shoe phone, Max spoke on his ...

- address book phone
- axe phone
- balloon phone
- belt phone
- briefcase phone
- Bunsen burner phone
- car radiator phone
- clock phone
- comb phone
- compact phone
- daisy phone
- donkey-shoe phone
- doughnut phone
- eyeglass phone
- fingernail phone
- fireplace phone
- freezer phone
- garter phone
- golf shoe phone
- gun phone
- hair dryer phone
- handkerchief phone
- headboard phone
- headlight phone
- hose phone
- hotline to the White House
- hydrant phone
- ice cream cone phone
- jacket sleeve phone
- lighter phone
- longhorns phone
- magazine phone
- microscope phone
- miniature phone in a normal pho
- mummy phone
- Agent 99's portrait phone
- perfume spray phone
- phone phone
- plant phone
- sandwich phone
- shepherd's staff phone
- sock phone
- steering wheel phone
- test tube phone
- thermos phone
- thumbnail phone
- tie phone
- wallet phone
- watch phone
- and water canteen phone.

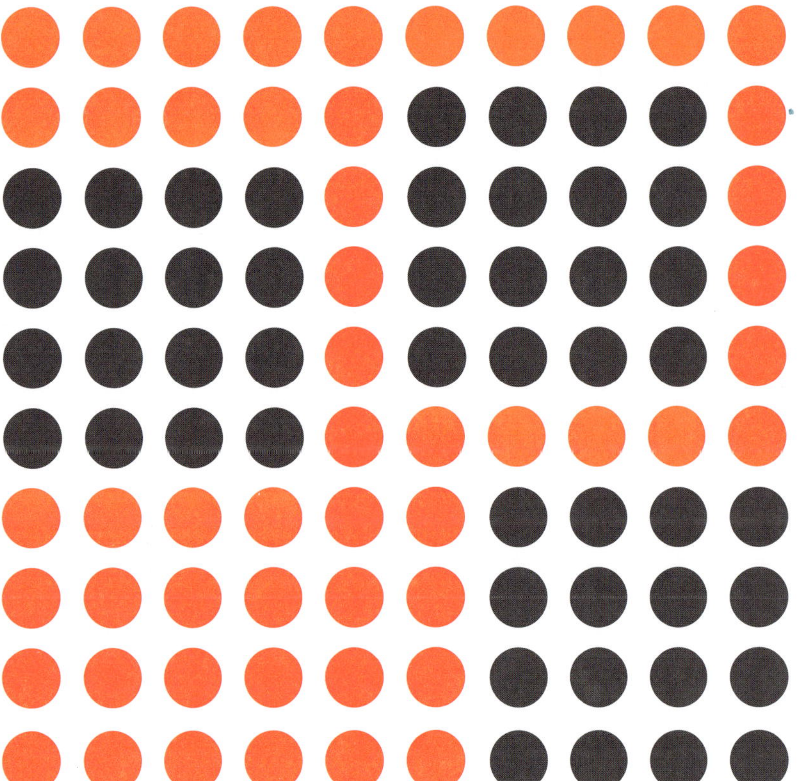

52

No matter how much you jumble a 15 square puzzle (see chapter 15), it can be solved in, at most, 52 moves.

There are 88 keys on a piano but they don't split 44 each between white and black. There are 52 white keys and only 36 black because in each octave while the white keys go from A to G, there are no black keys between B and C, nor E and F.

CAN'T TOUCH THIS

Fifty-two is an 'untouchable' number: it is never the sum of the proper divisors of any other number. What on Earth does that mean?

Well, the proper divisors of 38 are 1, 2 and 19:

and 1 + 2 + 19 = 22.

So '22 is the sum of the proper divisors of 38'.

So 22 is not an untouchable number.

Similarly, the divisors of 60 are 1, 2, 3, 4, 5, 6, 10, 12, 15, 20 and 30:

and 1 + 2 + 3 + 4 + 5 + 6 + 10 + 12 + 15 + 20 + 30 = 108.

So 108 is not an untouchable number.

But no number has a list of proper divisors which add up to 52, making 52 untouchable. The only other untouchable numbers under 100 are 2, 5, 88 and 96.

We know there is an infinite number of untouchable numbers, but whether there are any odd untouchables other than 5 remains an open (unsolved) problem.

The standard deck of cards used in the west has 52 cards, excluding jokers. The cards are adapted from the 56 numbered cards in the tarot deck, popular in medieval Italy. The design of the cards' faces was originally French, with the spades representing pikes (*piques*, in French) and the clubs the French trefoil symbol. The design of the back varies widely, from standard prints to advertising for whisky manufacturers via some truly low-grade pornographic photos. I'd strongly recommend that you inspect the back of any prospective pack before you utter the phrase, 'Happy birthday, Grandma!'

And for the record, a standard pack of 52 cards can be dealt out a whopping:

80,658,175,170,943,878,571,660,636,856,403,766,975,289,505,440,883,277,824,000,000,000,000

... ways.

52 PICK UP

Every 8 year old has encountered this moment of hilarity: take a deck of cards and ask a friend, 'Hey, wanna play this cool game I learnt at school today?'

Friend: 'Sure, what's it called?'

You: '52 pick up.'

Friend: 'How do you play?'

You: 'Well, there's 52 cards in a deck,' throwing the cards to the ground, 'pick 'em up. Hahahahahahahahahaha!'

Friend: often unprintable.

The game then sits unused by both parties for decades until one of them becomes a dad, tries it out on one of their kids and the tradition is reborn.

BELL NUMBER

Consider this problem: if you had 3 mobile phones of different colours, say Red, White and Purple, how many ways could you split them into groups?

There are 5 ways: Red and White together in one group and Purple by itself – let's write that as (RW)(P); there's (RP)(W), (WP)(R); there's all 3 phones in one big group (RWP); or we could put the 3 phones in separate groups (R)(W)(P).

To put that more mathematically, we say, 'There are 5 ways that 3 distinguishable objects can be arranged into non-empty sets'.

Eric Temple Bell thought about this sort of stuff a lot (objects and sets, not mobile phones. He died in 1960. Poor guy never even got to see a video recorder, let alone a mobile. 'Video recorder', you say? Ask your parents).

Thankfully, as the number of distinguishable objects gets larger, we don't need to work out all the ways one by one. Just 7 phones can be grouped 877 ways. Instead, the Bell numbers can be read off this neat little triangle:

1

1 2

2 3 5

5 7 10 15

15 20 27 37 52

52 ...

Each row begins with the next Bell number – which we get from the end of the previous row. To get the next number, you add the current number to the one above it to the right. Continue this until ... you've got something better to do.

So 52 is the fifth Bell number; or there are 52 ways to group 5 objects.

QUIZ QUESTION: The table above shows us that the fourth Bell number is 15. Find the 15 ways to arrange the 4 objects A, B, C and D into non-empty sets. ANSWER AT THE BACK OF THE BOOK

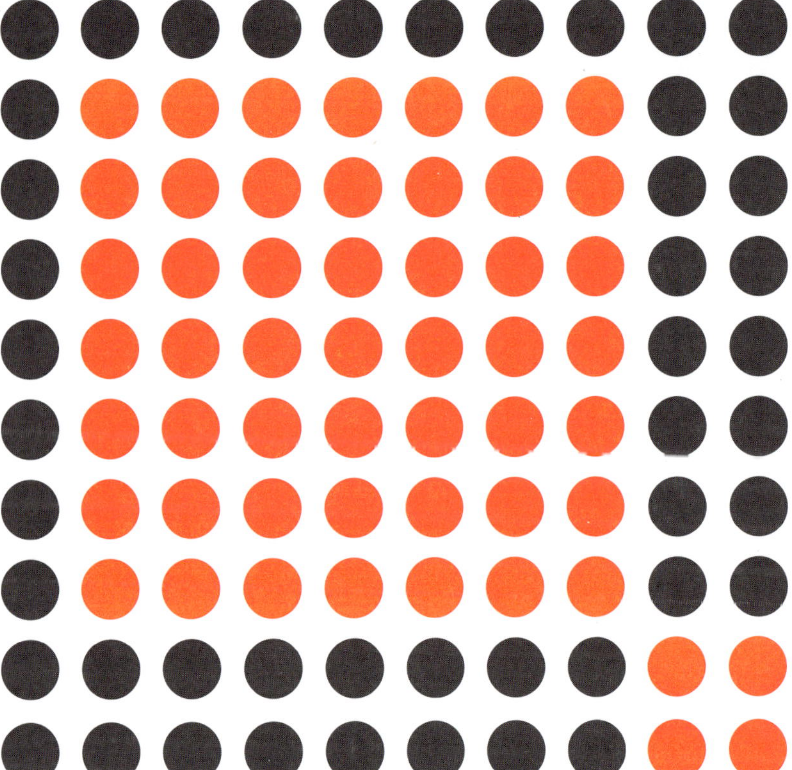

53

A PROBLEM

Answer this: 'Find the smallest positive number that, when you divide it by 3, leaves a remainder of 2 that, when you divide it by 5, leaves a remainder of 3 and that, when divided by 7, leaves a remainder of 4.'

Come on, let's have a go. If you want to do most of the work yourself, cover the page and just read one line at a time.

We'll start with the third requirement. What sort of number, when divided by 7, leaves a remainder of 4? Clearly 4 would do, as would 11, because it equals $7 + 4$, and 18, because it equals $14 + 4$.

The pattern is clear: a number that leaves a remainder of 4, when divided by 7, must be 7 times some number plus 4. To write a cool formula, x does the job if $x = 7k + 4$, where k is a whole number.

The other rules we need to satisfy are $x = 5m + 3$ and $x = 3p + 2$, where m and p are both whole numbers. (You can use k and m and p or anything you want. We just write different letters for each equation so we aren't suggesting that k, m and p have to be the same number. It's just the x that has to match up.)

So let's solve this tricky question by sheer hard work. The solutions to $x = 7k + 4$ are:

$$x = 4, 11, 18, 25, 32, 39, 46, 53, 60, 67, 74, 88, 95 \ldots$$

Write out the first few solutions to $x = 5m + 3$ and you should find only 3 numbers that match both lists. Those numbers are 18, 53 and 88.

Check the third rule as well by dividing each of these numbers by 3 and seeing what remainder you get. Not surprisingly, given which section we are in, the smallest positive answer that satisfies all 3 rules is ... 53.

53 of Saturn's 62 moons have names.

Saturn's 7 main moons are Mimas, Enceladus, Tethys, Dione, Rhea, Iapetus and Titan.

YOU, ME AND 53

Fifty-three is the smallest prime to have the 5 numbers either side of it composite, that is, none of 48, 49, 50, 51 and 52, nor 54, 55, 56, 57, and 58 are prime.

QUIZ QUESTION: Find the only other number under 100 for which this is the case. ANSWER AT THE BACK OF THE BOOK

The number 53 is also a 'self' number (see chapter 75) and, $53 = 2^2 + 7^2 = 1^2 + 4^2 + 6^2$.

SOPHIE GERMAIN PRIME

The number 53 is prime. In fact, it's a 'Sophie Germain prime' (see chapter 89 for the lowdown on Soph).

If you add up the first 53 primes you get 5830. Convince yourself of this, if you'd like, but remember that 1 is *not* prime.

But $5830 = 53 \times 110$ so the sum of the first 53 primes is divisible by 53.

This is not a very common occurrence.

QUIZ QUESTION: Apart from 1 and 53, find the only other number *n* less than 100 for which the sum of the first *n* prime numbers is divisible by *n*. BIG HINT: It's prime and it's in the 20s.
ANSWER AT THE BACK OF THE BOOK

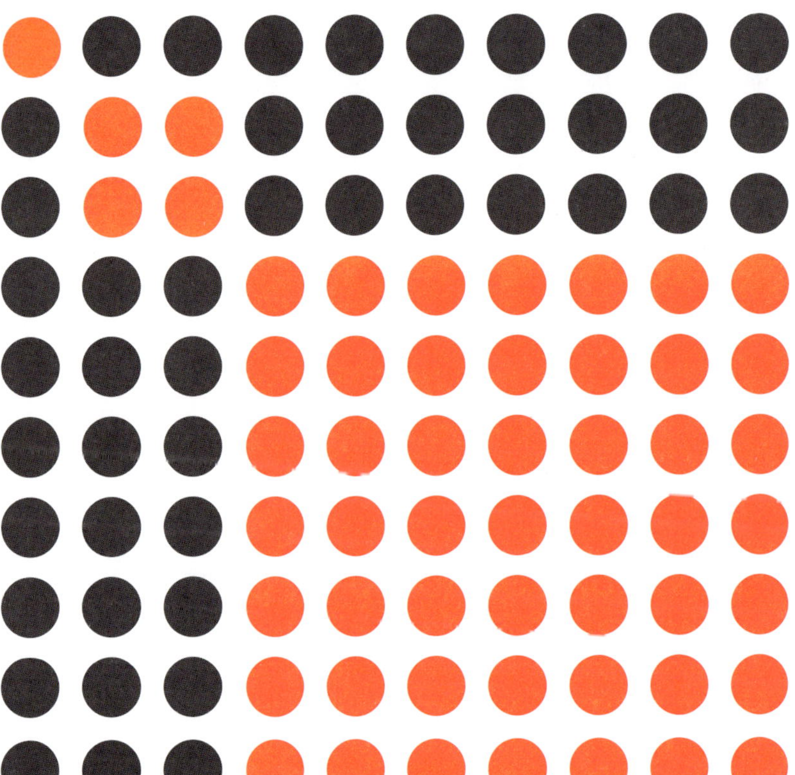

54

ABUNDANT

Fifty-four is an abundant number. See chapter 12 if you need a refresher on them.

STUDIO 54

Probably the most famous disco ever was Studio 54, at 254 West 54th Street in New York City. Over the dance floor, tubes studded with lights rose and descended. Welcome to the 1970s.

ZACK ATTACK

Zacharias Dase, born in 1824 in Germany, was an amazing human calculator who once deduced that:
$79{,}532{,}853 \times 93{,}758{,}479 = 7{,}456{,}879{,}327{,}810{,}587$
Time taken: 54 seconds.

UNIVERSAL QUADRATIC FORMS

The absolute gun mathematician Lagrange (born in Italy yet fiercely claimed by France where he did most of his work ... but that's another hornet's nest) showed that any 'positive whole' number can be written as the sum of 4 squares.

For example, $9 = 0^2 + 1^2 + 2^2 + 2^2$, $47 = 1^2 + 1^2 + 3^2 + 6^2$, $157 = 0^2 + 2^2 + 3^2 + 12^2$ and so on.

This is the same as saying every positive whole number can be written in the form $w^2 + x^2 + y^2 + z^2$ for some choice of whole numbers w, x, y and z.

It turns out that any positive whole number can be written in the form $w^2 + 2x^2 + 3y^2 + 4z^2$

So, $9 = 0^2 + 2 \times 1^2 + 3 \times 1^2 + 4 \times 1^2$, $47 = 3^2 + 2 \times 1^2 + 3 \times 0^2 + 4 \times 3^2$, and $157 = 1^2 + 2 \times 0^2 + 3 \times 2^2 + 4 \times 6^2$

Or as $w^2 + 2x^2 + 5y^2 + 8z^2$ ... and so on.

These combinations of w^2, x^2, y^2 and z^2 are examples of 'quadratic forms' and the brilliant Ramunujan showed that there are in fact 54 such 'universal quadratic forms'.

FAST FINGERS

There are 54 stickers on a Rubik's Cube. And, yes, peeling them off and putting them back on again in order is cheating.

Even if you do insist on removing the stickers, you probably won't solve the cube as quickly as the amazing Australian teenage cubemeister Feliks Zemdegs. At the age of 15, Feliks held an incredible 12 different cubing world records including the standard cube with 3 × 3 squares on each face but also the records for the 4 × 4 cube, 5 × 5 cube and 4 × 4 cube blindfolded!

He still holds the world one-handed record for solving a cube in 9.03 seconds.

At the 2011 Melbourne Open, Feliks set a personal best and then world record of 5.66 seconds ... yes, that's right, 5.66 seconds to solve a completely shuffled Rubik's Cube.

One person who *can* solve the cube faster is Mats Valk who broke Feliks' record in 2013 at the Zonhoven Open in Belgium with a time of 5.55 seconds. But I've met Feliks and he's pretty awesome so I really wanted to include him in this book.

Another thing that I love about the Australian cubing community is that they parody the Australian horseracing public holiday, Melbourne Cup Day, with an annual Melbourne *Cube* Day. In the unlikely event I ever become Prime Minister of Australia, I will immediately declare Melbourne Cube Day a public holiday (which is one of the many reasons I'll never be PM!)

But it's not just human beings that are solving Rubik's Cubes at incredible speeds. Before you go Googling 'Cat solves cube' just wait up a bit. Instead, Google 'Cubestormer' – a robot built by David Gilday and Mike Dobson using a Lego Mindstorms customisable robot kit and a Samsung Galaxy S4.

In March 2014 at the Birmingham Big Bang Fair, Cubestormer 3 solved a Rubik's Cube in ... wait for it ... seriously are you sitting down ... 3.253 seconds.

PICTURE SEMIPERFECT

We've met the perfect numbers in chapters 6 and 28. A number is perfect if it is equal to the sum of all of its divisors except for itself. So, the factors of 28 are 1, 2, 4, 7, 14 and 28 and $1 + 2 + 4 + 7 + 14 = 28$ making 28 a perfect number.

If a number is equal to the sum of some of its factors but not all of them, the number is called 'semiperfect'. It turns out that every multiple of a perfect or semiperfect number is ... semiperfect.

So, because we know that 6 is perfect and $54 = 9 \times 6$, we know that 54 is semiperfect. Indeed, the divisors of 54 are 1, 2, 3, 6, 9, 18, 27 and 54 and 54 can be written as the sum of some of its divisors, namely $54 = 3 + 6 + 18 + 27$.

Don't be offended, 54, there's nothing wrong with being semiperfect. It's a lot closer than most of us get.

LEYLAND PRIMES

We can write $54 = 3^3 + 3^3$ which makes it a Leyland number (a number which can be written in the form $x^y + y^x$, x and y integers greater than 1). The Leyland numbers up to 100 are 8, 17, 32, 54, 57 and 100. You might like to calculate their forms $x^y + y^x$; then again, you may not.

YOU WISH ...

To birdie every hole on a standard par 72 golf course is considered a perfect score. It would mean getting around in 54 strokes. The fact that in professional golf no-one has ever shot under 58 proves this is a fairly tough definition of perfection.

QUIZ QUESTION: 54 is the smallest number that can be written as the sum of three squares in three different ways. The squares don't all have to be different in each sum. Find them. ANSWER AT THE BACK OF THE BOOK

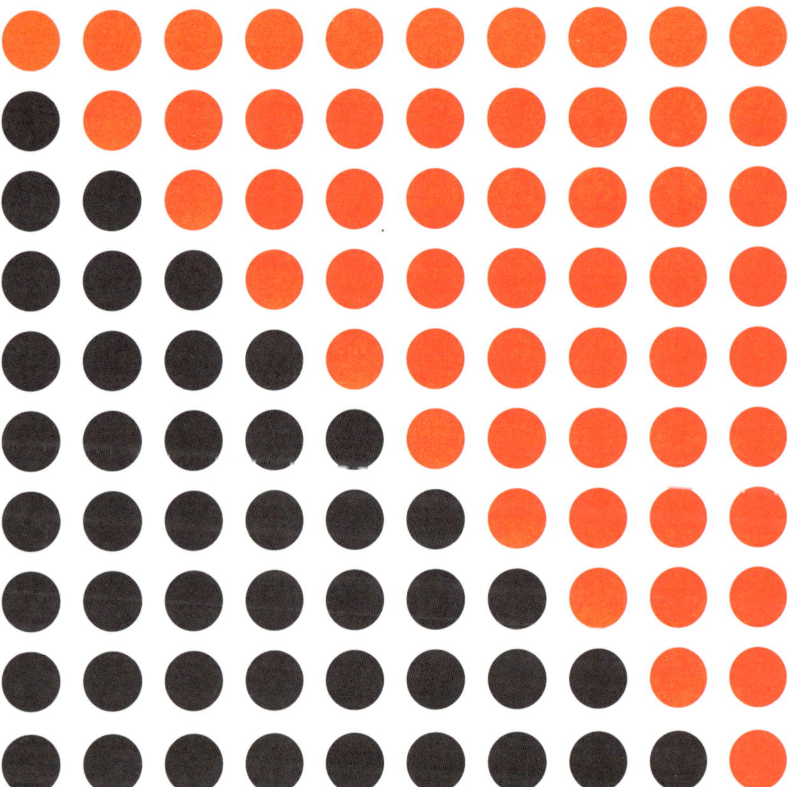

55

PYRAMIDAL BALLS

Try stacking a pile of tennis balls so each layer is a square.

Pretty difficult, hey? But if you could, the top layer would have 1 ball, the second 4 balls, the third 9 balls, then 16 and so on. The total number of balls are the 'square pyramidal' numbers: 1, 5, 14, 30, 55 ... So 55 is the fifth square pyramidal number.

QUIZ QUESTION: What are the next 3 square pyramidal numbers after 55? ANSWER AT THE BACK OF THE BOOK

MY FAVOURITE TRIANGULAR, KAPREKAR, SQUARE PYRAMIDAL FIBONACCI NUMBER, EVER

Fifty-five is also the 10th Fibonacci number, the 4th Kaprekar number (see chapter 45) and the sum of all the numbers from 1 to 10. So:

$$1 + 2 + 3 + 4 + 5 + 6 + 7 + 8 + 9 + 10 = 55$$

... which makes 55 the 10th triangular number. If you really want to intimidate someone, you could say, 'What's your favourite triangular, Kaprekar, square pyramidal, Fibonacci number? Mine would have to be 55.' If they suggest any other example they're bluffing! Apart from 1, the number 55 is the only such example.

DISTINCT SQUARES

Once you get above the number 128 every number can be written as the sum of 'distinct squares'. For example, $135 = 1^2 + 3^2 + 5^2 + 6^2 + 8^2$.

For smaller numbers this is harder to do because you don't have as many possible squares to use. In fact, 28 of the numbers from 1 to 100 can't be written as distinct squares:

2, 3, 6, 7, 8, 11, 12, 15, 18, 19, 22, 23, 24, 27, 28, 31, 32, 33, 43, 44, 47, 48, 60, 67, 72, 76, 92, 96.

The number 65 has a very cute property to do with distinct squares that you'll meet shortly, but before we get there ...

QUIZ QUESTION: Of the numbers less than 100 that can be written as distinct squares, 55 and 88 are the only ones that require more than 4 distinct squares. Write 55 and 88 as the sum of distinct squares.

ANSWER AT THE BACK OF THE BOOK

EDWIN SHARPE AND THE MAGNETIC ZEROS 1 DIRECTION KRS-ONE 1 G
LEAP ONE DAY AS A LION CHECK 1-2 PLAYER ONE KONONO NO1 BELL X1
2 UNLIMITED U2 2PAC 2 LITRE DOLBY 2 DOOR CINEMA CLUB BOYZIIMEN 2
CREW 2 IN A ROOM H2O SISTER2SISTER REEL 2 REEL US3 3OH3 ALABAMA 3
DIRTY THREE 3 DOG NIGHT 3 THE HARD WAY TIMBUK 3 SPACEMAN 3 FUNB
3 DAYS GRACE 3 DOORS DOWN 3 COLOURS RED 3 MILE SMILE 3RD EYE B
MOJAVE 3 THREE DEGREES OPUS 3 THANE RUSSAL AND THREE STEVE WYNN
THE MIRACLE 3 4 NON BLONDES AC 4 THE KLEIN 4 MEGA CITY 4 4 TOPS FOUR
THE 4 KINSMEN 4PM DILLINGER FOUR GANG OF FOUR COSMO4 LAB4 UNIT F
PLUS TWO JURASSIC 5 MAROON 5 MC5 5 FINGER DEATH PUNCH BEN FOLDS
STAR 5IVE JACKSON 5 5 SOS DAVE CLARK 5 THE 5,6, 7,8S 5 FOR FIGHTING HI5
ZICATO FIVE COUNT 5 GRANDMASTER FLASH AND THE FURIOUS 5 DEADM
LOUIS ARMSTRONG AND HIS HOT FIVE BR5-49 FIVE MAN ELECTRICAL BAND
KING AND THE FIVE STRINGS FIVE MILE TOWN DELTA 5 CATEGORY 5 THE SUC
MAN 5 ELECTRIC SIX 6 FEET UNDER SIX FINGER SATELLITE 6 AND OUT SIX F
HICK THE SCARAMANGA 6 SIXPENCE NONE THE RICHER APPOLONIA 6 VANI
SLANT 6 BLUE SIX AREA 7 ZERO 7 S CLUB 7 SHED 7 LEVEL 7 L7 AVENGED 7 F
THE MAX WEINBERG 7 7DUST 7OF9 7MARY3 SCHOOL OF SEVEN BELLS LOUIS A
STRONG AND HIS HOT SEVEN 8 FOOT SATIVA EIGHTH WONDER SK8 OR DIE
TER8 DV8 THE HOT 8 BRASS BAND BUTTER 08 CLUB 8 GRADE 8 LISA LOEB AI
STORIES NINE INCH NAILS BUCK OH 9 RADIO 9 STS9 GICHY DAN'S BEECHW
#9 10CC 10 YEARS AFTER 10 MINUTE WARNING 10 OUTTA 10 10 TENORS FIN
ELEVEN ELEVEN STELLA ONE ELEVEN 12 FOOT NINJA 13TH FLOOR ELEVATORS LC
XIV M16 HEAVEN 17 N17 EXCUSE 17 EAST 17 18 VISIONS MATCHBOX 20 SECTA
CIDA SIGLO 20 CATCH 22 SYNC24 APARTMENT 26 28 DAYS 30 SECONDS TO M
30 ODD FOOT OF GRUNT JONAH 33 36 CRAZYFISTS E37 (NOW ENGIN3) .38 L
E-40 SUM 41 LEVEL 42 +44 JUNE OF 44 BOCA 45 BLACK 47 49ERS 50 CENT E
OL' 55 59 TIMES THE PAIN STARFLYER 59 THE DEAD 60S SUN 60 65 DAYS OF ST
EIFFEL 65 BRASIL 66 67 SPECIAL 69 EYES SHAM 69 PINK CREAM 69 FELA KUTI
AFRICA 70 MEXICO 70 SR-71 JJ72 THE DELTA 72 PREFUSE 73 78 SAAB LINK 80 GOT
ATHLETICO SPIZZ 80 TAHITI 80 MX-80 SOUND BUMBLEBEEZ 81 M83 THE 88 ROC
88 CURRENT 93 ROUTE 94 OLD 97S 98 DEGREES 99TH FLOOR ELEVATORS HAIRCUT
HIGHWAY 101 BLINK 182 CORPORATION 187 ISOTOPE 217 THINKING FELLERS UN
LOCAL 282 311 MU330 SWIRL 360 THE 411 APOLLO 440 GALAXY 500 BR 549 808 ST
MC 900 FT. JESUS URSULA 1000 SPOT 1019 1927 DEATH FROM ABOVE 1979 TER
2000 BRAN VAN 3000 THE KELLEY DEAL 6000 GO!GO!7188 10 000 MANIACS INFI

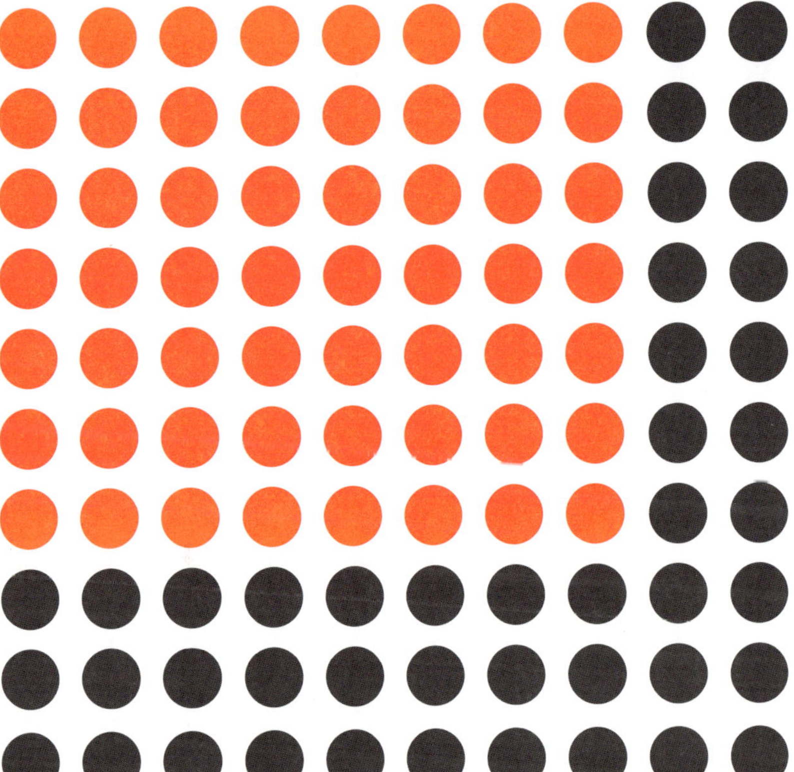

56

Joe DiMaggio about to kiss his signature baseball bat.
Source: Wikipedia

Between 15 May and 17 July 1941, Joe DiMaggio ascended to baseball immortality with his famous 56-game hitting streak.

Martial arts fighter Frank W. Dux also claims a record hitting streak of 56 consecutive knockouts in a single tournament.

Dux's claim is shrouded in controversy but Frank, if you're reading this ... having seen the movie *Bloodsport* where the Muscles from Brussels, Jean-Claude Van Damme, plays you ... I'm down with anything you say, buddy!

The 'Aubrey holes' are a ring of 56 chalk pits dating to the earliest phases of the construction of Stonehenge in the late 4th century BC. Named after John Aubrey, they quite possibly served an astronomical purpose.

TETRAHEDRONS

If you stack tennis balls in triangular layers, the layers will consist of 1, 3, 6, and 10 balls and so on. A triangular pyramid is a tetrahedron, so the tetrahedral numbers are 1, 4, 10, 20 ... You add the number of balls in each layer together to get the numbers in the series, so:

$1 + 3 = 4$

$1 + 3 + 6 = 10$

$1 + 3 + 6 + 10 = 20$ ... and so on.

Can you see that the tetrahedral numbers are just the sums of the triangular numbers? Similarly, looking back at chapter 55, you'll see that the square pyramidal numbers are the sum of the square numbers.

QUIZ QUESTION: Convince yourself that 56 is the 6th tetrahedral number. ANSWER AT THE BACK OF THE BOOK

BREAKING Ba

One of my all-time favourite television shows *Breaking Bad* also had a very strong geeky side. The lead character Walter White was a high school chemistry teacher and the show revolves around chemistry, so in the opening credits various letters in people's names are enlarged to read as chemical symbols.

So in *Breaking Bad* the Br becomes the symbol for bromine and Ba for barium.

For many people, this is the closest they've actually come to the symbols for barium and bromine, but what do these graphics actually mean?

Barium was discovered by one of the giants of chemistry Humphrey Davy in 1808 and the diagram tells us that:

137.33 +2

Ba

56

2-8-18-7

137.33 is the atomic mass of barium.

56 is the atomic number of barium – that is there are 56 protons in the nucleus of an atom of barium and as you read across a periodic table, barium comes up as the 56th element.

The +2 refers to something called the oxidation state.

The line 2-8-18-7 is the electron shell configuration, describing how many electrons are in each orbit around the nucleus.

But if you're swotting for a chemistry test don't use the cover of a *Breaking Bad* DVD as your textbook.

The electron shell configuration for barium is wrong, it's just bromine's copied (see how 2 + 8 + 18 + 7 doesn't equal 56). This is probably because it looked better that way and the producers rightly didn't expect anyone to study for their university finals off a *Breaking Bad* T-shirt!

Barium's configuration is 2-8-18-18-8-2 making it one of only 8 elements with a palindromic electron shell configuration. Look up a periodic table to find the other 7.

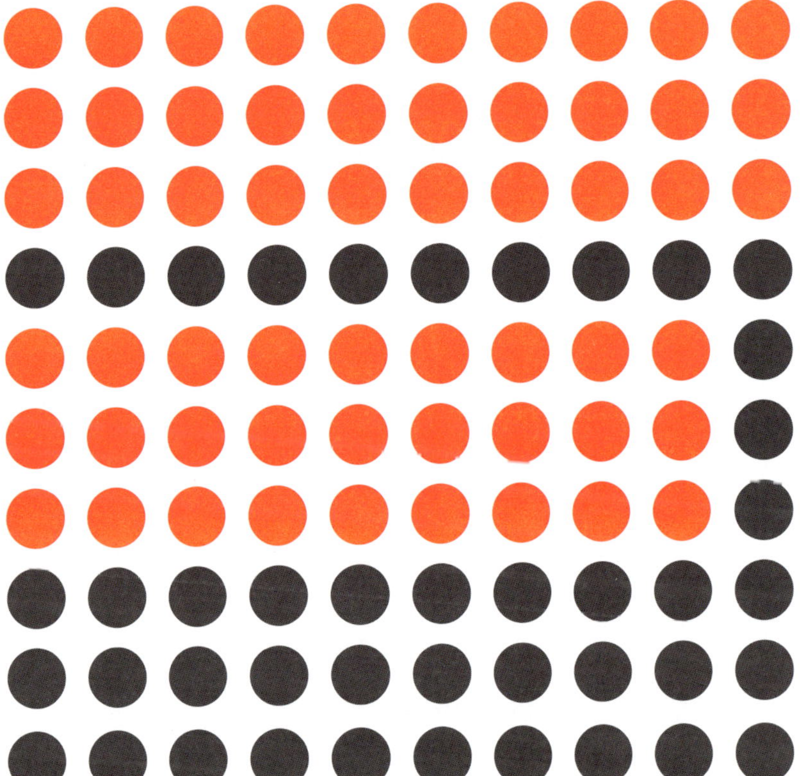

57

OODLES OF NOODLES

In China, calligraphy is revered as an artform. Mastering the sublime world of Chinese character writing takes years of practice and requires proficiency with the highest quality implements. In many ways, writing Chinese characters is seen to be as much of an artform as painting.

So the character below is particularly special. It symbolises a type of noodle popular in the Shaanxi Province – the BiangBiang noodle.

'Biang' is one of the most complex Chinese characters in contemporary usage, and requires 57 brush strokes.

The Australian Southern Giant Darner Dragonfly is the fastest insect that's been reliably measured. Its top speed is a blistering 57 kilometres an hour.

QUIZ QUESTION: $57 = 2^5 + 5^2$ and hence is a Leyland number. Find all the numbers from 1 to 100 that can be written in the form $a^b + b^a$ with both a and b not equal to 1. ANSWER AT THE BACK OF THE BOOK

HEINZ 57

In New York in the 1890s, German-American chef and entrepreneur Henry Heinz coined one of advertising's most famous catchphrases to describe the wide range of ketchups, sauces and relishes he'd created. The slogan '57 varieties' was a big hit, despite the fact that Heinz actually produced over 65 products at the time. Henry allegedly liked the look of 57.

PAINTING FACES OF A CUBE

If you take a blank white cube and paint each face either red, yellow or blue, there are 57 distinct ways you could label the cube so that picking them up and spinning them around they still looked different.

ANOTHER WAY AROUND A CIRCLE

The first way we learn to measure angles is in degrees, with 180 degrees in a straight line and therefore 360 degrees in a circle. But once we move into calculus and start to analyse functions like sin x, cos x and so on, we tend to measure angles in a new unit called 'radians'.

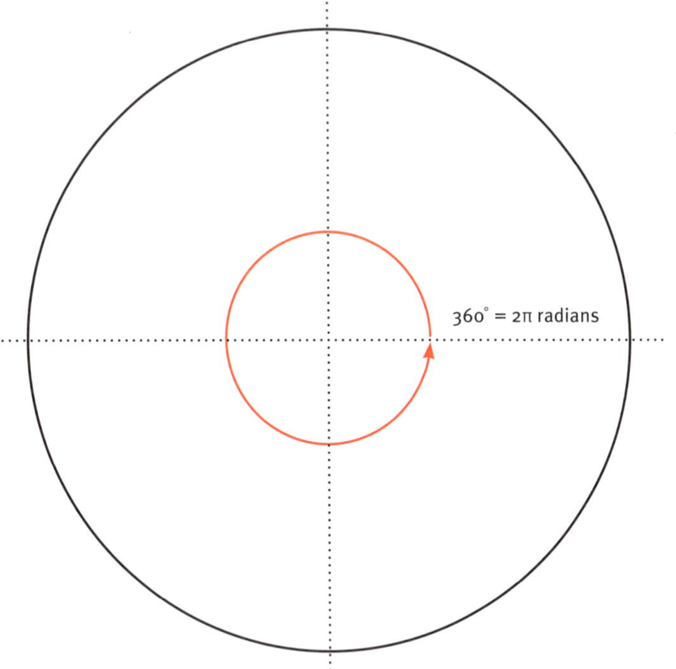

There are 2π radians in a circle (for reasons I won't go into here, but they are pretty cool reasons). To compare degrees to radians we have:

2π radians = 360 degrees

So 1 radian = 360 ÷ 2π degrees
= very close to 57.3 degrees.

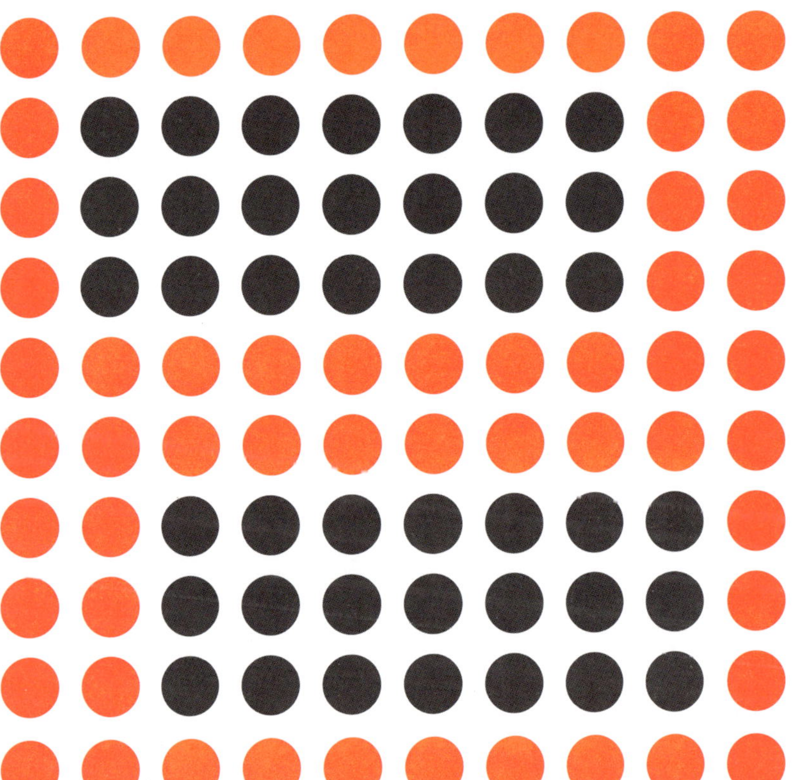

58

The sum of the first 7 prime numbers is 58.

The planet Mercury's orbit around the Sun isn't circular. It ranges from 47 to 70 million kilometres and averages just under 58 million kilometres. It hurtles through space at 35 kilometres per second, the fastest of all of our solar system's planets, though Earth is no slouch at almost 30 kilometres per second.

SWORD SWALLOWERS RULE, OKAY?

Natasha Veruschka holds the world record for the longest sword swallowed which measured 58 centimetres (22.83 inches) on Sword Swallowers Awareness Day, 28 February 2009. I'm not sure what social grievance Sword Swallowers Day is aiming to address but I sincerely hope it's achieving its aims.

DIE HARD FAN

58 Minutes is a thriller by American author Walter Wager on which the film *Die Hard 2* was based. If Walter was still alive, I'd love to shake his hand and thank him on behalf of my wife who is obsessed with the *Die Hard* franchise – at least up to the fourth instalment: *Live Free or Die Hard* (2007). To be honest, the 2013 offering, *A Good Day to Die Hard* ... well, she could take it or leave it.

NUMBER GAMES

Let's play a game with the number 58.

Squaring its digits and adding them together, 58 becomes:

$5^2 + 8^2 = 89$

Now do the same to 89 and it becomes:

$8^2 + 9^2 = 145$

Even though 145 has three digits, it's not hard to do the same process again and get:

$1^2 + 4^2 + 5^2 = 42$

Keep going and you'll get the pattern:

58, 89, 145, 42, 20, 4, 16, 37, 58, 89, 145, 42 ...

... which obviously goes forever.

You should also be able to see that if I started on any other number in this loop, I'd stay in the loop. For example, 16 becomes $1^2 + 6^2 = 37$,

37 becomes $3^2 + 7^2 = 58$
58 becomes 89 and so on.

Just tuck this away in the back of your mind and you'll encounter it again soon.

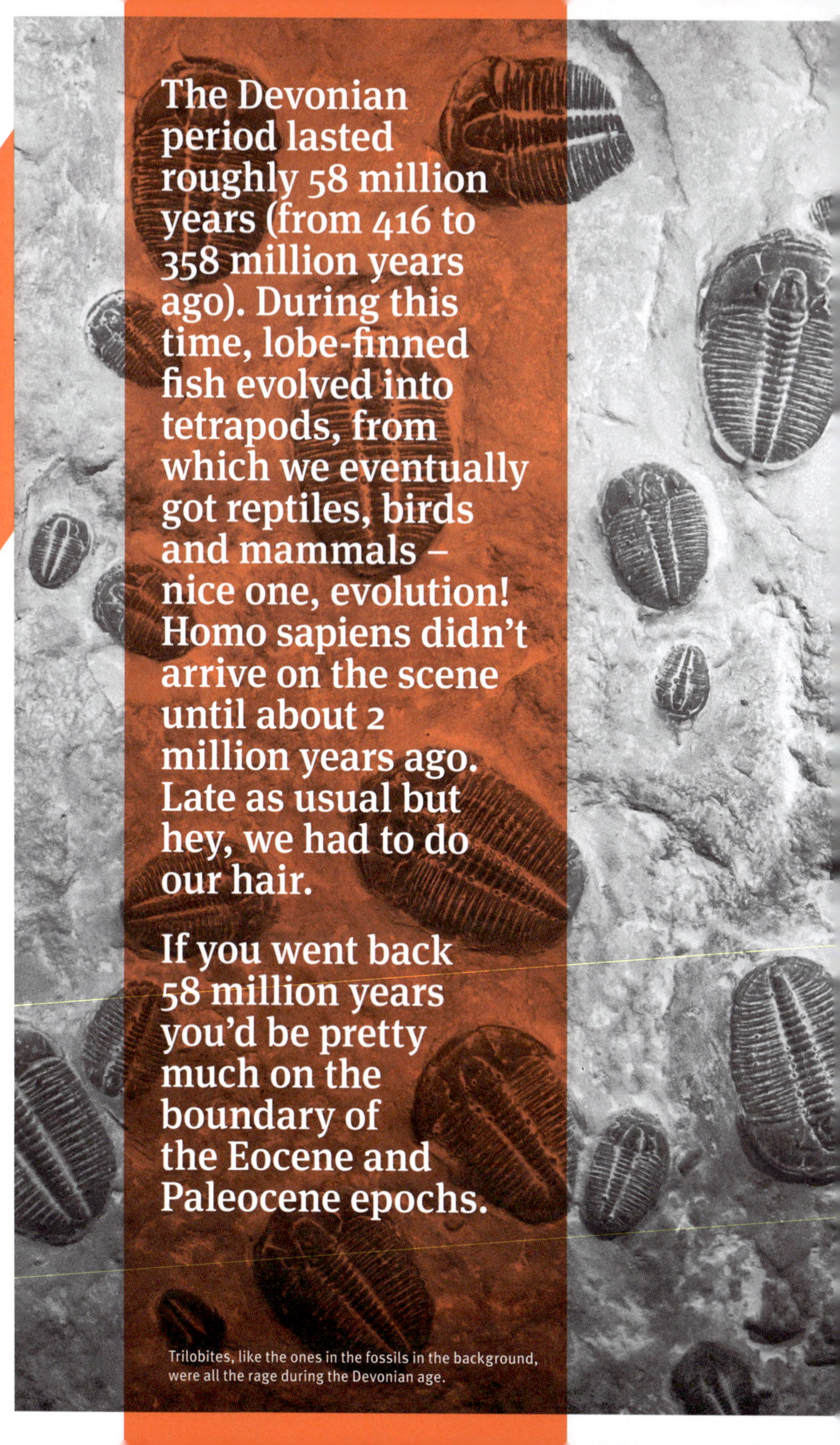

The Devonian period lasted roughly 58 million years (from 416 to 358 million years ago). During this time, lobe-finned fish evolved into tetrapods, from which we eventually got reptiles, birds and mammals – nice one, evolution! Homo sapiens didn't arrive on the scene until about 2 million years ago. Late as usual but hey, we had to do our hair.

If you went back 58 million years you'd be pretty much on the boundary of the Eocene and Paleocene epochs.

Trilobites, like the ones in the fossils in the background, were all the rage during the Devonian age.

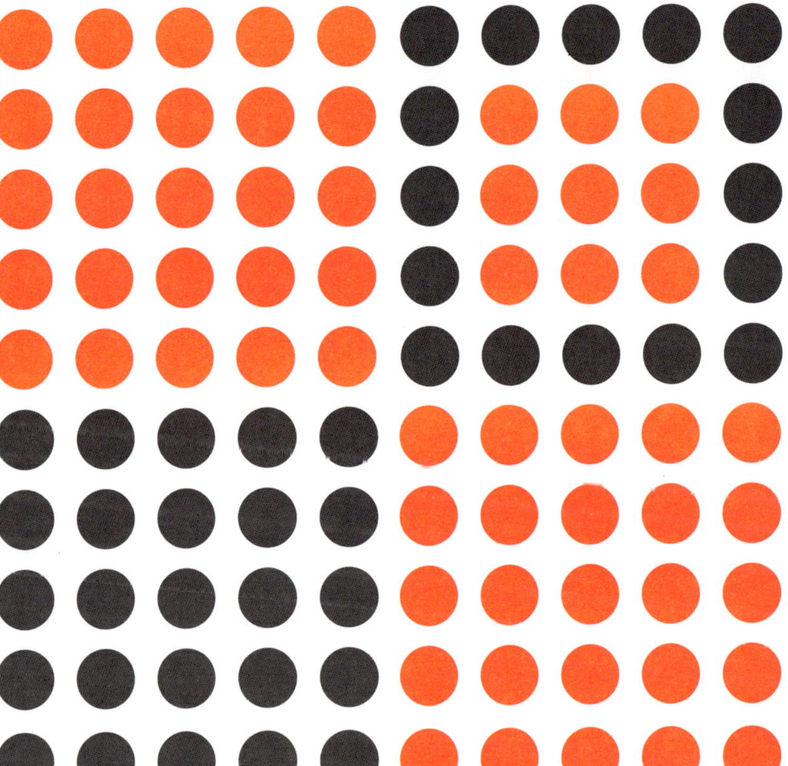

59

DOGGONE

A dog has 59 leg bones on each side – 29 in the foreleg and 30 in the hindleg ... and ideally one in its mouth.

A 15-ROUND

boxing match lasts – you guessed it: 59 minutes.

SHOOT A 59

To shoot a round of 59 in golf is considered a holy grail of sorts. Al Geiberger was the first man to do so on the PGA tour and in 2001 the great Annika Sörenstam became the first woman to break the 60 mark.

Baseball pitcher Orel Hershiser holds the record for the most consecutive scoreless innings thrown in the US major leagues. Orel threw down 59 no-scorers in a row which, depending on your point of view, is either the most amazing achievement or the most interminably boring sporting record imaginable. No offence, Orel.

ODE TO EULER

Leonhard Euler proved that:

$$635{,}318{,}657 = 59^4 + 158^4 = 133^4 + 134^4$$

That is, there is a number that is the sum of two fourth powers in two different ways. And yes, 635,318,657 is the smallest such number.

If you stop to think for a second how you could possibly discover this using only ink and paper, you get a momentary glimpse into the monumental genius that was Leonhard Euler. #MyMathematicalMancrush

HUMAN CANNONBALLS

Until I started researching this book I never knew the role Australians played in the history of human cannonballs. For some reason they passed right over that when I was in high school.

But it turns out in Sydney back in 1872 Ella Zuila and George Loyal grabbed a special piece of circusing fame: George was shot out of a cannon and Ella caught him while hanging from a trapeze bar.

Just stop and read that again a second time. Yes ... cannon ... caught ... hanging ... trapeze!

Well, 140 years later, the two indomitable names in the community of Human Cannonballs are David Smith ... and David Smith!

David Smith Sr held the world record for longest human cannonball flight until 2011 when his son, David 'The Bullet' Smith Jr broke it flying 59.05 metres from the mouth of the cannon to the furthest point on the impact net.

But just when you thought this couldn't get any more exciting, some people claim David Smith Sr still holds the record with a 61-metre propulsion set a decade before his son's mighty flight.

Don't tell me there isn't a movie in this – someone get Will Ferrell's agent on the line!

I'm going to leave this dispute to David and David to sort out at one of what I can only imagine are their fairly intense Christmas lunches.

SO NEAR AND YET SO FAR ...

If one day you had some spare time and thought, 'Hey, I'll just multiply the primes together and add 1 and see what I get' you'd find that:

$2 \times 3 + 1 = 7,$

$2 \times 3 \times 5 + 1 = 31,$

$2 \times 3 \times 5 \times 7 + 1 = 211$

and

$2 \times 3 \times 5 \times 7 \times 11 + 1 = 2311$

All of which are prime numbers. But just as a frisson of excitement was coming over you and you were thinking, 'Here we go – I'm about to get a famous theorem named after me' you'd discover that:

$2 \times 3 \times 5 \times 7 \times 11 \times 13 + 1 = 30{,}031$ which equals 59×509

Trust me from personal experience – you'll recover.

In fact, we'll soon see in chapter 83 that multiplying some prime numbers together and adding one to the result allows us to understand one of the simplest but most important results concerning prime numbers. It's all to do with a famous theorem by one of the great mathematicians of the ancient world, Euclid.

If you really want, you can jump ahead to it right now – it's a beautiful little theorem. But when you've finished, make sure you come back to chapter 60 and read from here. The next few chapters are absolute belters.

09

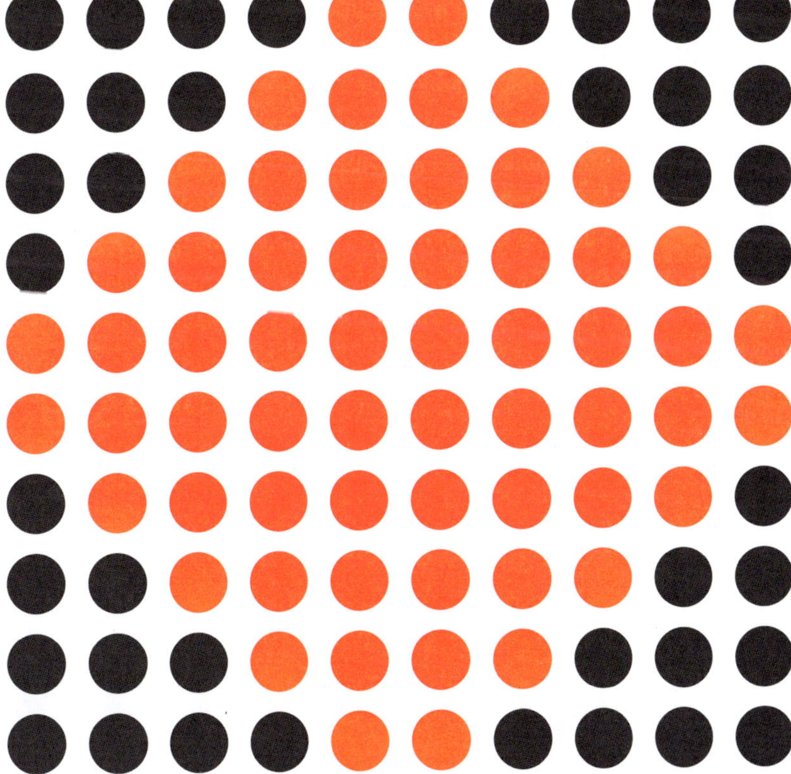

BASE-60

The Babylonians and Chaldeans counted in base-60 (see chapter 10 if you've forgotten what that means). This is hard to get our base-10 heads around, nevertheless a base-60 system is called a 'sexagesimal' system. But 60 has the advantage of being easily divided by 2, 3, 4, 5 and 6 as well as 10, 12, 15, 20 and 30. The Babylonians divided a circle into 360 degrees and today we still divide hours into 60 minutes and minutes into 60 seconds. And you guessed it, they came up with some very cool calculations involving the number 60.

For example, when calculating the Pell numbers (check out chapters 29 and 70) we see that $\frac{41}{29}$ is an approximation of $\sqrt{2}$.

A famous diagram on a clay tablet almost 4000 years old shows that mathematicians in the Old Babylonian period worked out an approximation of $\sqrt{2}$ (which is 1.4142135...) given by:

$$1 + \frac{24}{60} + \frac{51}{60^2} + \frac{10}{60^3} = \frac{30547}{21600} = 1.4142196\ldots$$ Pretty close, hey!

Similarly, the Greek mathematician Hipparchus estimated a year as having $6 \times 60 + 5 + \frac{14}{60} + \frac{44}{60^2} + \frac{51}{60^3}$ days. This estimate of 365.24579... days compares to the time it takes for the Earth to revolve around the Sun (a sidereal year) of on average 365.25636... days.

And another Greek, Ptolemy, came up with an approximation for pi (3.141593...) in base-60 (or sexagesimal) of:

$$\pi = 3 + \frac{8}{60} + \frac{30}{60^2} = 3.141666\ldots$$ Nice work, P-dog.

BUCKYBALL

Carbon is one of the most pervasive of all elements. It is the fourth most abundant element after hydrogen, helium and oxygen, and is called the basis of all life. Carbon can be found in all lifeforms and about $\frac{1}{5}$ of all your body is carbon.

One of the coolest ways that carbon occurs, is in giant molecules called 'buckyballs'.

Below is a diagram of C_{60}, a molecule of 60 carbon atoms. Look at C_{60} with its 32 faces – 20 of them are hexagons (with 6 atoms around the edge) and 12 are pentagons (with 5 atoms around the edge). This shape should be familiar to you ... hint it looks pretty much like a common soccer ball design. If you're thinking truncated icosahedron, well done you. If not, pop back to chapter 13 for a second.

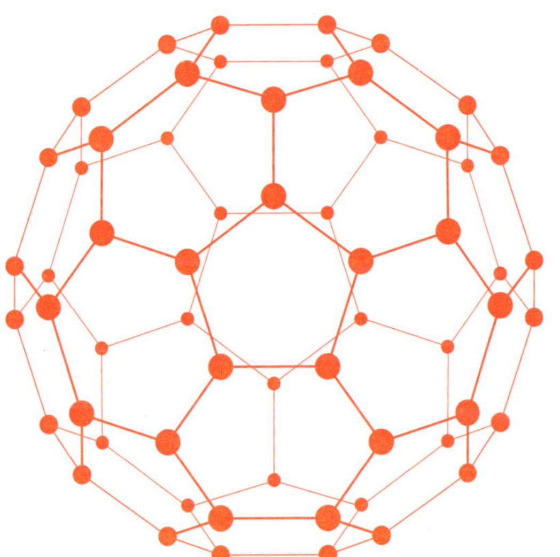

There are also buckyballs C_{70}, C_{76} and C_{84} and they all have 12 pentagonal faces – only the number of hexagonal faces vary.

If you want to know another way to arrange atoms of carbon, check out graphene – cool!

Sixty is another abundant number and 60 degrees is the interior angle of an equilateral triangle.

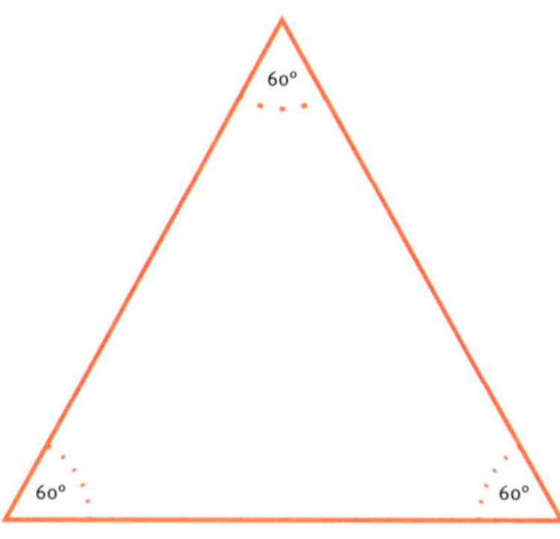

AN ICOSIDODECAHEDRON

... has 32 faces, 30 vertices and 60 edges. Check that Euler's formula $V + F - E = 2$ applies to this Archimedean solid and while you're at it notice that it can be cobbled together from the faces of an icosahedron and a dodecahedron.

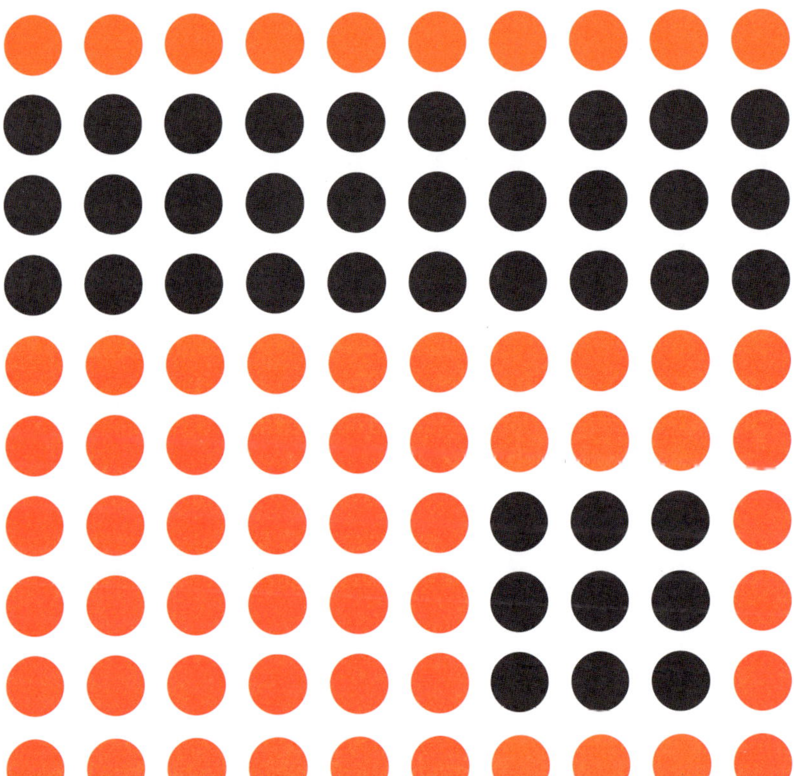

61

DIABOLICAL DIOPHANTINES

Back in the 1700s, they very rarely had the gigantic week-long parties that we call mathematics conferences. Similarly, there weren't many mathematical journals or magazines. So the main way of communicating your ideas and achievements was by writing to another mathematician.

One day, Fermat banged out a letter to his old pal l'Hôpital, challenging him to find whole numbers x and y which would solve:

$$x^2 - 61y^2 = 1.$$

An equation like this, where the solutions are whole numbers, is called a 'Diophantine' equation (see chapter 84), and Fermat ate such equations for breakfast.

Euler used to call this Pell's equation, which is bizarre because Pell had nothing to do with it – but that's another story for another time. Anyway, I won't make your head explode by trying to solve it; the smallest answer is:

$$x = 1{,}766{,}319{,}049 \text{ and } y = 226{,}153{,}980$$

Hey, you think that's tough? Fermat also worked out that the smallest whole-number solution to $x^2 - 109y^2 = 1$ is:

$$x = 158{,}070{,}671{,}986{,}249 \text{ and } y = 15{,}140{,}424{,}455{,}100$$

QUIZ QUESTION: Try to solve $x^2 - 37y^2 = 1$, with the hint that $x = 73$.

ANSWER AT THE BACK OF THE BOOK

PUZZLING PENTOMINOES

We met the 12 pentominoes in chapters 12 and 28 and we've seen interesting ways to fit polyminoes together. Well, the largest rectangular fence we can put around a shape using the 12 pentominoes is the 11 × 11 square below which encloses an area of 61 unit squares.

Can you work out how we use the 12 pentominoes to make the fence?

If we don't insist on the fence being 4 straight sides we can enclose the 128 unit squares below. Can you create this fence from the 12 pentominoes? You can check your answers at the back of the book.

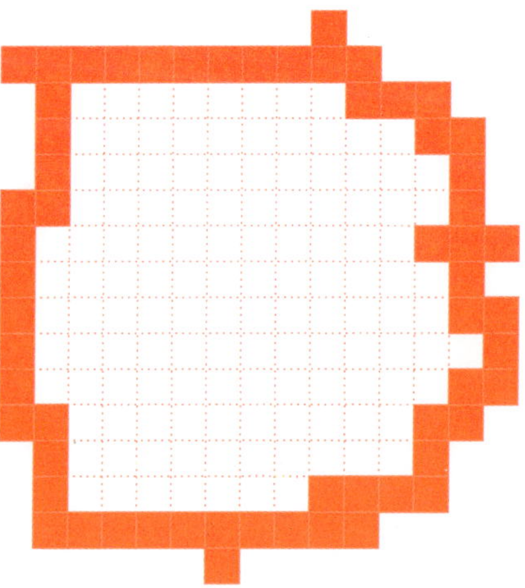

A

S T

A N D

A R D S

N E L L E

N E Y E S I

G H T C H A R

T H A S S I X T

Y O N E L E T T

E R S O V E R E

L E V E N R O W S

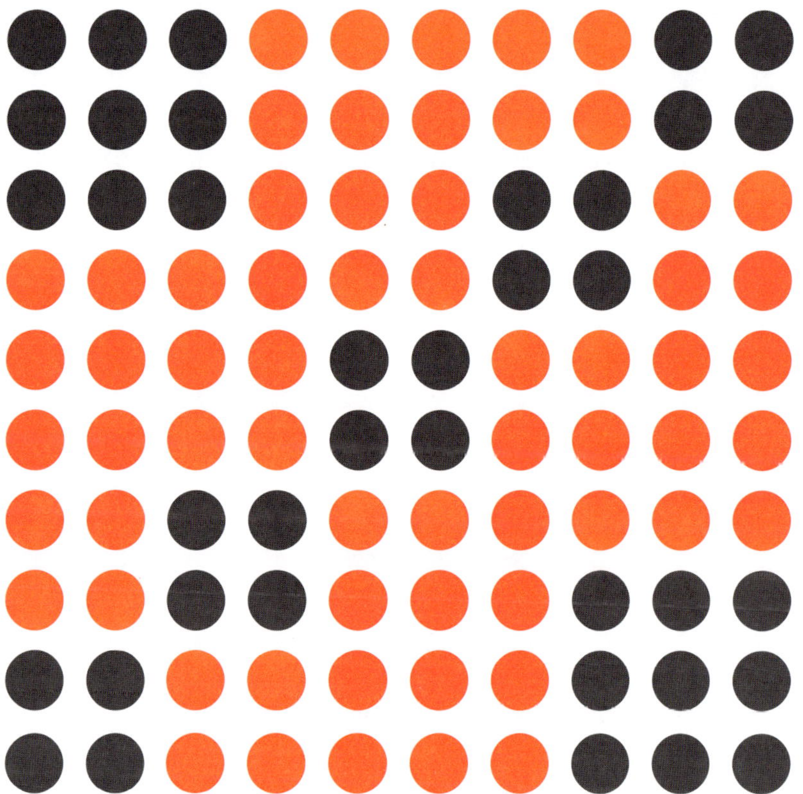

62

DOMINOES

There are $64 = 8 \times 8$ squares on a chessboard. Now, remove the two opposite corners like this:

QUIZ QUESTION: Can you cover the remaining 62 squares with 31 standard dominoes?

HINT: Take a chessboard, place a coin on the two corners to rule them out and try to fit the 31 dominoes on. Also try removing the coins and just placing the 31 dominoes down however you can – what do you always notice about the two squares that aren't covered? ANSWER AT THE BACK OF THE BOOK

A GREAT RHOMBICOSIDODECAHEDRON

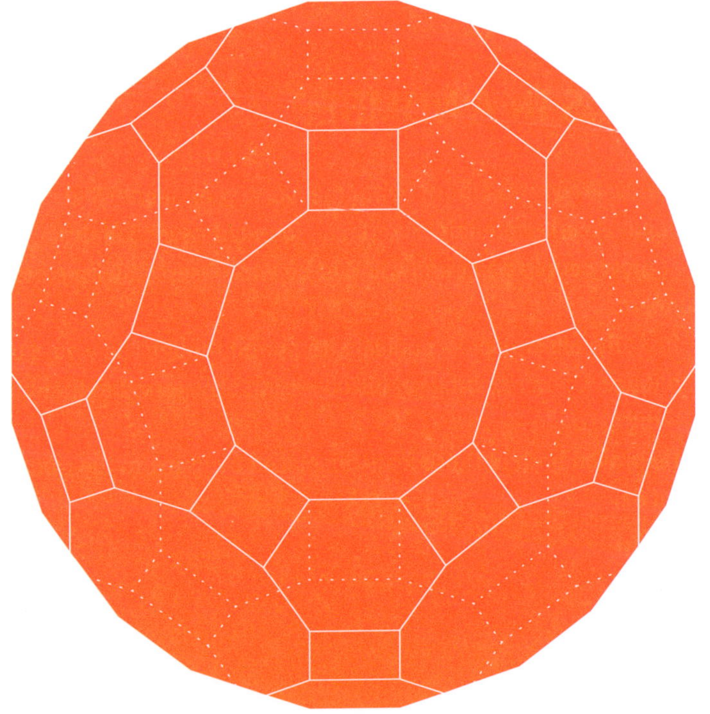

... has 62 faces, 120 vertices and 180 edges. Check that Euler's formula $V + F - E = 2$ applies to this, the largest of all Archimedean solids.

GIDDY UP

The oldest known horse died at the age of 62. He was named, appropriately, Old Billy.

THAR SHE BLOWS

When wind speeds get above 62 kilometres per hour, the world meteorological organisation upgrades a deep depression to a cyclonic storm which, in Australia, we call a category 1 cyclone.

VELVET BROWN

Sixty-two was Velvet Brown's winning raffle ticket number in the 1944 film *National Velvet*.

WHAT'S IN A NUMBER

Sigmund Freud was born with the very cool name Sigismund Schlomo Freud. More importantly, he founded psychoanalysis, but what most people don't know is that he was obsessed with certain numbers.

Initially, Sig was preoccupied with 51 and thought he would die at that age. Upon reaching 52, his obsession transferred to thoughts of dying at 62.

He proved both of those fears misplaced by living to the ripe old age of 83.

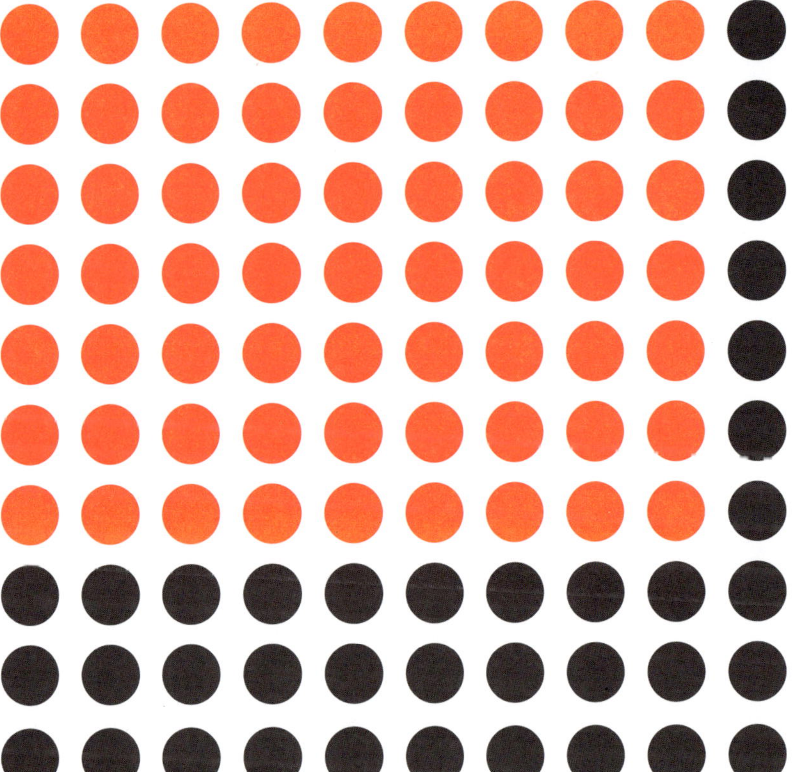

63

WWW ... WTF

While most of us dread long domain names and regularly have to say over the phone, 'Just wait a sec, I'll get a pen', Subrahmanyam Karuturi loves welcoming people to:

http://iamtheproudownerofthelongestlongestlongestdomainnameinthisworld.com

This site was registered in mid-2005, but as far back as 02 the domain:

thelongestdomainnameintheworldandthensomeandthensomemoreandmore.com

... was floating around. These both contain 63 letters which is as long as a .com domain name can get. When asked to confirm a world record for registering the world's longest computer domain name, the *Guinness Book of World Records* declined, saying such a record had 'no merit' to it – and was 'similar to taking the largest number in the world and adding 1'.

Of the myriad 63-letter domain names my favourite is a page that has had over 2 million visitors and gives pi to 1 million places – I'm talking about:

3.141592653589793238462643383279502884197169399375105820974944592.com

On the topic of ridiculously long domain names, Google once famously erected billboards that read:

{first 10-digit prime found in consecutive digits of *e*}.com.

The first such 10-digit prime when you read out the expansion of the mathematical constant *e* is 7,427,466,391. People who went to 7427466391.com got another maths problem and so on. It was a recruiting campaign for maths megabrains.

The Greek Archimedes

is thought by many to be the greatest mathematician of the ancient world and one of the mightiest mathematical minds ever. Among other things, he loved calculating massive numbers. In his book *The Sand Reckoner*, he says the universe would be big enough to hold:
8×10^{63} grains of sand, which is 8 followed by 63 zeros, or 8,000,000,000,000,000,000,000,000,000,000,000,000,000,000,000,000,000,000,000,000,000,000,000.

This is called not 8 million, or 8 billion, but 8 *vigintillion*. Though when I say 'called' we very rarely have cause to use a word for a number with 63 zeros. In fact, unless you're describing Archimedes' estimate for the size of the Universe in terms of grains of sand, this might be the last time you see it for a while, so let it roll off your tongue. 'Vigintillion ... vigintillion ... vigintillion ...'

'Archimedes Thoughtful' painted by Domenico Fetti in 1620.
Source: Public Domain

IS ...

the age at which,on New Year's Eve, 1990 at Covent Garden, in the UK, the great Australian coloratura soprano Dame Joan Sutherland exercised her sublime voice for its last public performance.

the number of chromosomes found in the offspring of a donkey and a horse.

and also the number of days for which both cats and dogs are pregnant – although seldom to each other, despite the best efforts of many *Funniest Home Videos* contestants.

And of course the least impressive mountain in the world and our lowest named peak is Mount Alvernia in the Bahamas, which towers a mere 63 metres above sea level. I mention this intending no disrespect to those mighty adventurers who have conquered Mount Alvernia.

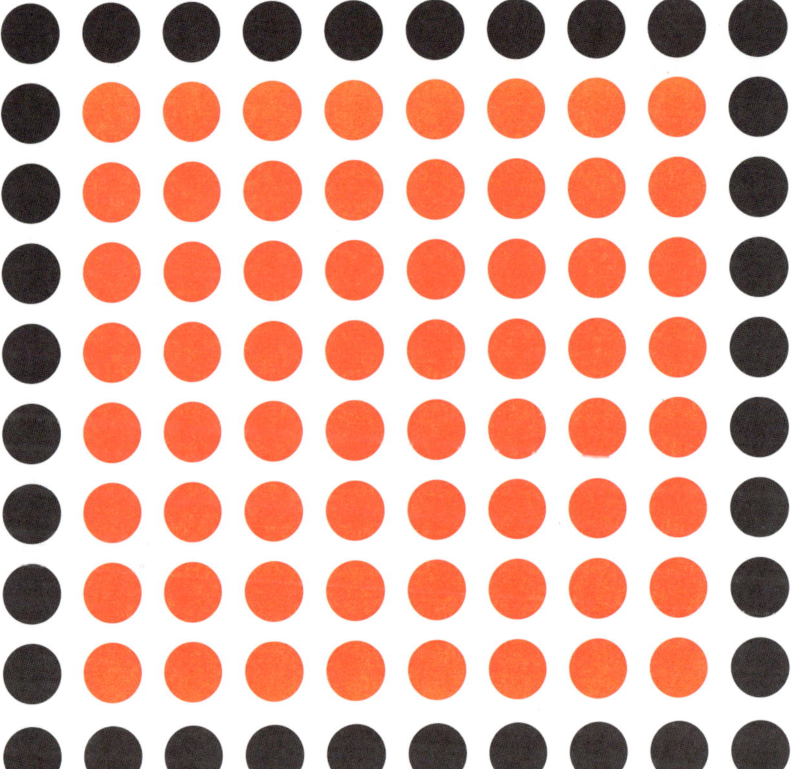

64

POWER TO 64

Sixty-four is the sixth power of 2 and 6 is 2×3 so 64 can be written as three different powers: $64 = 8^2 = 4^3 = 2^6$.

Next time you want to impress your pals, tell them that 64 is really 1,000,000. They'll stare at you in bewilderment until you explain 'binary numbers'. We've already met base-10 (see chapter 10) – well, binary numbers use base-2.

We write 7038 the way we do because it is $7 \times 10^3 + 0 \times 10^2 + 3 \times 10 + 8 \times 1$. Each digit explains how many powers of 10 we need to give the number we want.

So to write a number in binary, we break it down not into powers of 10, but into powers of 2. For example, $25 = 1 \times 16 + 1 \times 8 + 0 \times 4 + 0 \times 2 + 1$

Thus, in binary numbers, we would write this as 11,001.

Now, 64 is a power of 2, so to write it in base-2, or binary form, we see that: $64 = 1 \times 64 + 0 \times 32 + 0 \times 16 + 0 \times 8 + 0 \times 4 + 0 \times 2 + 0 \times 1$

Therefore, 64 = 1,000,000 in binary.

BEN FRANKLIN'S MAGIC SQUARE

Benjamin Franklin is probably most famous for being one of America's Founding Fathers but in fact he did an incredible number of things – at the very highest level. He was America's diplomat to both France and Sweden, first Postmaster general, and the President of Pennsylvania. Aside from all that, he was a successful inventor, scientist, author and newspaper founder. He profoundly influenced the country America is today.

As a scientist and inventor, he made breakthroughs in electricity including the lightning rod, bifocals, and the Franklin stove, among others. Hey Ben, just settle down and leave something for everyone else to do!

He also didn't mind the occasional mathematical doodling – hey who doesn't? In a letter to a friend in 1769 he described an 8×8

magic square that he had discovered some time previous. Check out the numbers in the square below:

52	61	4	13	20	29	36	45
14	3	62	51	46	35	30	19
53	60	5	12	21	28	37	44
11	6	59	54	43	38	27	22
55	58	7	10	23	26	39	42
9	8	57	56	41	40	25	24
50	63	2	15	18	31	34	47
16	1	64	49	48	33	32	17

The constant for an 8 × 8 magic square has to be (1 + 2 + 3 + 4 + 5 + 6 + 7 + 8 + 9 + 10 + 11 + 12 + 13 + 14 + 15 + 16 + 17 + 18 + 19 + 20 + 21 + 22 + 23 + 24 + 25 + ... up to 64) ÷ 8 = 260 (for a shortcut in adding up 1 + 2 + ... + 64 see chapters 21, 81 and 100). But not only does each row and column add up to 260; starting at any edge square and adding along you see that every half row and every half column add to 130 (half of 260); every smaller 2 × 2 square adds up to 130; the four corners add up to 130, and if you come in one square towards the centre those four squares add up to 130 ... and so on. Keep looking at this square and you will find more hidden beauties. Well played, Mr Franklin.

The marvellous Beatles song 'When I'm Sixty Four' was written by Paul McCartney and released in 1967 on their album *Sgt. Pepper's Lonely Hearts Club Band.*

Although the theme is ageing, it was one of the first songs McCartney wrote, when he was only 16.

I'm willing to be corrected, but I'd bet it's also the most famous pop song featuring a 2 soprano, 1 bass clarinet trio.

The great Bobby Fischer, who, ironically died at the age of 64, faces off against t 'totally awesome but not quite as great' World Champion Mikhail Tal at the Leipzig Olympiad 1 November 1960. Source: Deutsches Bundesarchiv

There are 64 squares on a chess-board

Sexy chess words include 'chaturanga' and 'shatranj' – Indian board games from which chess probably evolved; 'Ruy Lopez' and 'anti Veresov variation' – different possible opening tactics; and, my favourite, 'zugzwang' – a situation where you are forced to move, even though any move you make will hurt you. At many times during its history, chess was banned by religious leaders and royalty, including King Louis IX, who outlawed the game in France in 1254. How this could have been a priority while Europe battled Mongol hordes, the Crusades raged and France was at war with the English Plantagenet kings eludes me.

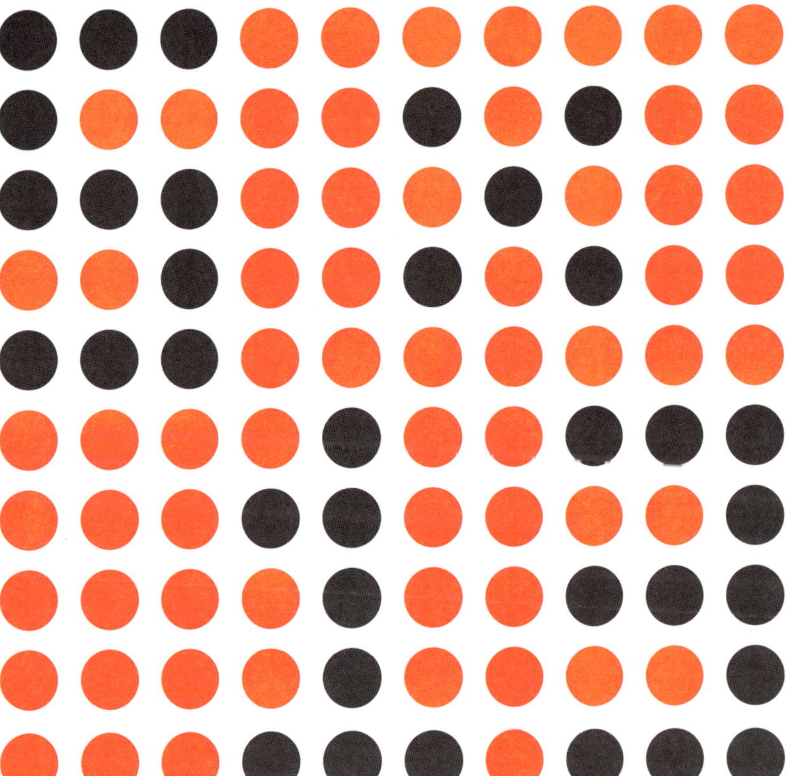

65

MORE MAGIC

Sixty-five is the constant of a 5 × 5 magic square. Maybe I should explain why.

The square contains all the numbers from 1 to 25, so if we added up all the squares we'd get:

1 + 2 + 3 + 4 + 5 + 6 + 7 + 8 + 9 + 10 + 11 + 12 + 13 + 14 + 15 + 16 + 17 + 18 + 19 + 20 + 21 + 22 + 23 + 24 + 25 = 325 (for the quick way to add up numbers like this see chapter 21, 81 or 100).

But each square has 5 rows and they all add up to the same so each row must add up to 325 ÷ 5 = 65.

Here's are a couple of 5 × 5 magic squares:

1	7	23	20	14
18	15	4	6	22
9	21	17	13	5
12	3	10	24	16
25	19	11	2	8

1	10	22	18	14
17	13	4	6	25
9	21	20	12	3
15	2	8	24	16
23	19	11	5	7

And here's four I haven't quite finished yet. You know the drill – they get harder.

1		19	13	7
	8	2	21	
22	16	15	9	3
10		23	17	
18	12			24

17	24		8	
23	5			16
	6		20	
10			21	3
11	18		2	9

23				15
	18	1	14	22
17	5		21	9
	12	25	8	16
	24	7	20	

25	13	1	19	7
16	9	22	15	3
4	17	10	23	11

Answers, as always, are at the back of the book.

I'M OUT OF HERE

Sixty-five is probably the most popular retirement age around the world. But with the increasingly ageing population in most countries, that's all about to change.

In Australia, for example, the retirement age was set at 65 in 1908, but in 1908 someone who was 15 years old could be expected to live until the ripe old age of ... 64.

Retirement was only meant to happen in the 'unusual' case of someone exceeding their life expectancy. A hundred years on, with people regularly living until 85 or longer, something has to give. Countries around the world are approaching the very politically sensitive issue of increasing their retirement ages for fear of their pension systems collapsing and sending them broke.

ALL POWER TO 65

The number 65 is 'semiprime' and part of the 'Padovan' sequence, both of which you'll meet in chapter 86. It's also 'octagonal', which you won't meet until chapter 96.

But when it comes to powers 65 really shines.

Firstly, and I'd never seen this before until about three days before I had to have this book finished and I'm so happy I saw it because for me this sort of stuff is beautiful:

$$65 = 1^5 + 2^4 + 3^3 + 4^2 + 5^1$$

Ain't that sweet!

Also $65^7 = 1414^3 + 2{,}213{,}459^2$

I know, I know, you're thinking, 'For crying out loud, Adam, why did you go and ruin my day by showing me that?' But this is one of only a handful of equations connected to the Fermat-Catalan conjecture you'll meet in chapter 71.

If you'd really like to push yourself, I'll tell you that:

$$65^7 = 4{,}902{,}227{,}890{,}625$$

Now take a pen and paper (I know, old school) and with a bit of good old-fashioned long multiplication and grunt, convince yourself that $1414^3 + 2{,}213{,}459^2$ equals the same.

And to top it all off ...

QUIZ QUESTION: Sixty-five is the smallest number that can be written as the sum of two distinct positive squares two different ways. How? Too easy? Okay, 65 squared can be written as the sum of two distinct squares in four different ways. Find all four! ANSWER AT THE BACK OF THE BOOK

99

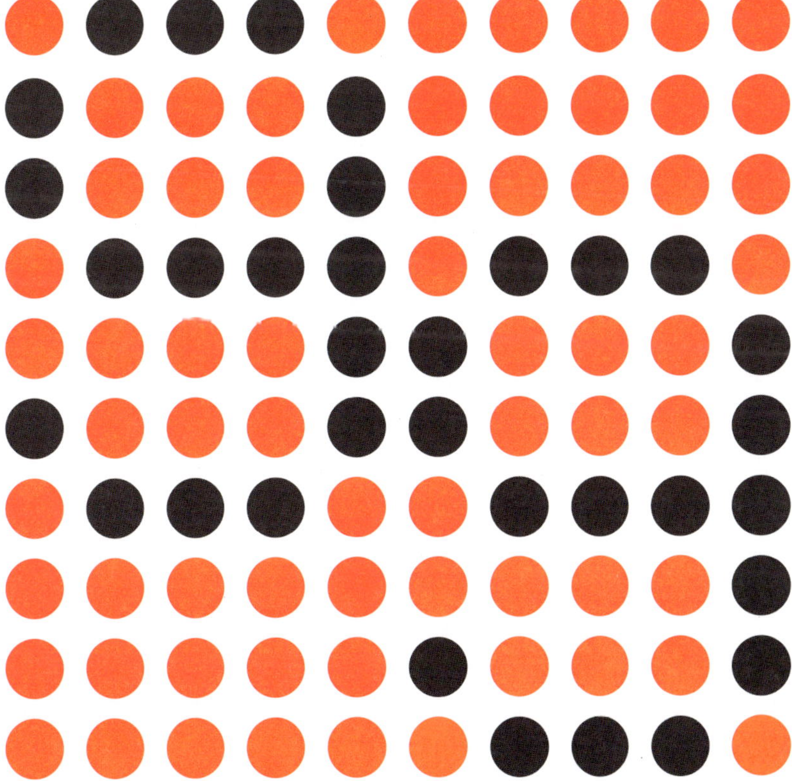

PALINDROMIC AND TRIANGULAR

Sixty-six is the next triangular number after 55 (see chapter 21). It's also palindromic, which means it reads the same way backwards as it does forwards.

QUIZ QUESTIONS: Find the next two palindromic triangular numbers after 66. And show that 66 is an abundant number (see chapter 12).

ANSWERS AT THE BACK OF THE BOOK

GET YOUR KICKS

Route 66, also known as 'The Highway of America', was one of the original roads in America's highway system and was opened in 1926. It runs about 2500 miles (4000 kilometres) from Chicago to Santa Monica but the better known description of its path features in the song 'Get your kicks on Route 66':

Well it goes from St. Louis, Joplin, Missouri
Oklahoma City looks ooh so pretty
You'll see Amarillo and Gallup, New Mexico
Flagstaff, Arizona don't forget Winona
Kingsman, Barstaw, San Bernadino

While the song is known for its many versions by everyone from Nat King Cole and Chuck Berry, to The Rolling Stones and Depeche Mode, it was actually written by Bobby Troup in 1946 after he drove for 10 days west to Hollywood seeking songwriting fame.

Route 66 isn't the only appearance of 66 in modern music. Jazz Latino groovemaster Sergio Mendes and his band Brasil 66 have been churning out his bossa-nova funk gear since the 1960s and are still going strong.

GOING DOTTY

Look closely at this diagram and you'll notice that I've drawn 5 dots and using only two colours I've joined up all the pairs of dots without forming any triangle that is all the one colour.

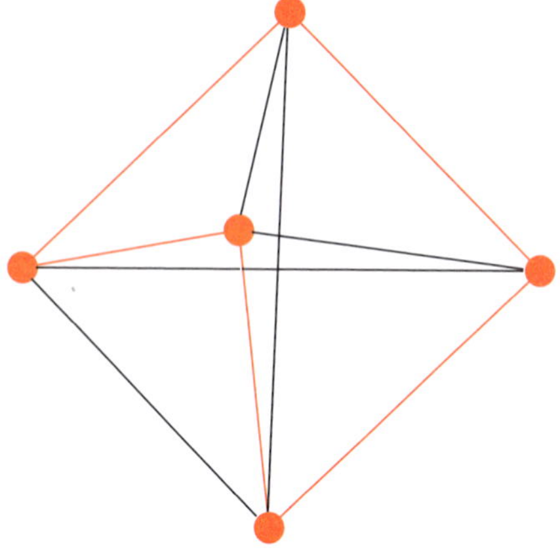

You can try this yourself with 5 random dots and 2 different coloured pens.

Now draw 6 random dots and try it. Keep trying until you get really irritated ... and then stop.

For any collection of 6 dots and 2 colours when you join up all pairs of dots you have to have made a triangle all the one colour.

If we have three different coloured pens, you can join 16 dots without forming a triangle (it's not easy but you can) but once you have 17 dots, any way you join up the pairs will create a triangle of one colour.

If you have 66 dots on a page and 4 different coloured pens, any joining of all the pairs of dots has to form a triangle with all three sides the same colour. The proof of this is (comparatively) simple. But 66 is the last 'upper bound' that can be shown fairly simply.

We have in fact pushed the number down to 62 dots for the 4 colour problem, but the proofs are so difficult and esoteric that even a complete nerd like me starts to feel a little queasy reading them!

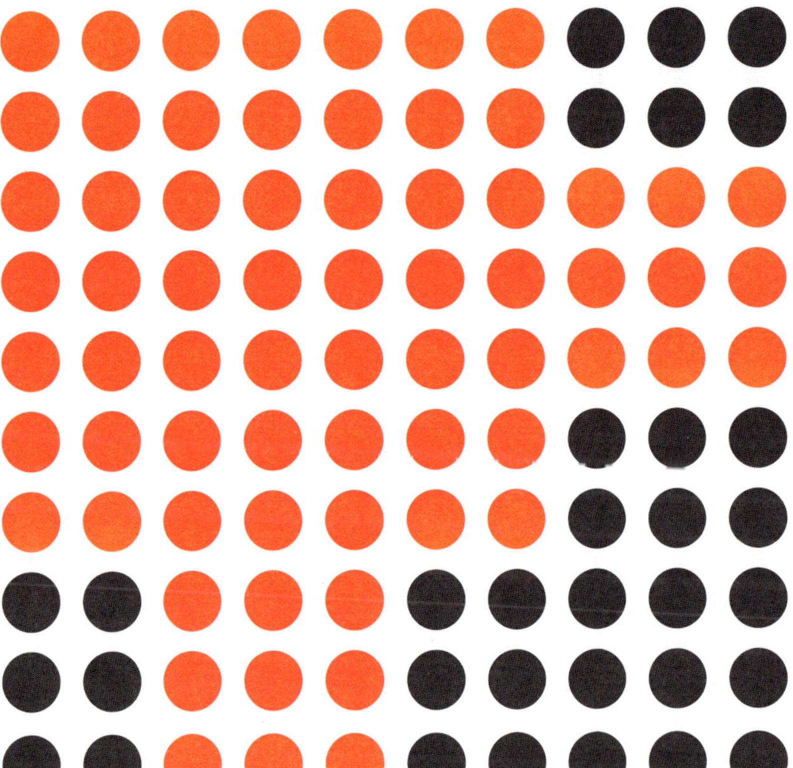

67

PIERRE DE FERMAT'S LAST THEOREM

As we've already seen, it was known as early as 500 BC that in a right-angled triangle (a triangle with a 90-degree angle) the lengths of the sides are related by the Pythagorean Equation $a^2 + b^2 = c^2$.

Over 2000 years later, Frenchman Pierre de Fermat scribbled a thought in the margin of a textbook of complicated things.

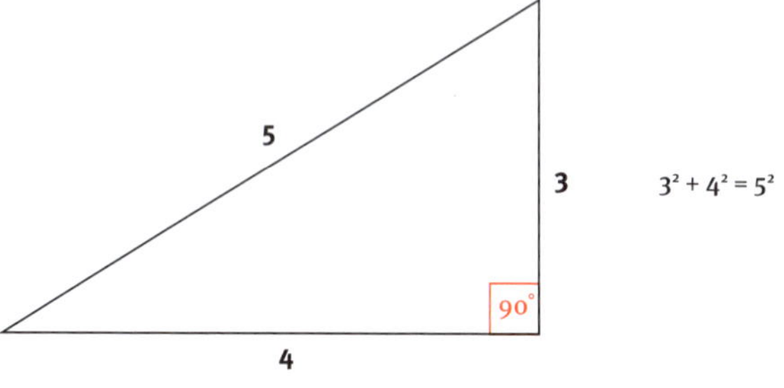

Fermat claimed that, while there are an infinite number of examples of $a^2 + b^2 = c^2$, like the one that we see above, there are no whole numbers that will satisfy $a^3 + b^3 = c^3$ or $a^4 + b^4 = c^4$ or $a^{31} + b^{31} = c^{31}$. Indeed, there's no solution to $a^n + b^n = c^n$ for whole numbers if n is greater than 2.

Fermat reckoned he could prove this, but there wasn't enough room in the margin of the book to write out his proof. While he might have thought that he could, he almost certainly could not. The problem baffled the greatest minds for centuries and only succumbed to a monstrously long proof by British mathematician Andrew Wiles – 358 years after Fermat first raised the idea!

For a long time, though, mathematicians had suspected that the claim was true and had pushed the limits higher and higher for the lowest possible value of n if there was to be a counterexample. Basically, they knew that $a^{5,000,000} + b^{5,000,000} = c^{5,000,000}$ couldn't happen, nor could it for any powers less than 5,000,000, but it took ages to bust the problem for all possible values of n.

Why am I telling you this? Well, one of the first major players trying to solve Fermat's Last Theorem (as this problem was known) was German megabrain Ernst Edward Kummer. He proved that there was no solution to $a^n + b^n = c^n$, when n was a special type of prime that he called regular (which, despite sounding straightforward, has a wicked definition that I won't go into here). When he'd done that, all the cases from $n = 1$ to $n = 100$ could be instantly removed, except 3 curly, irregular primes. They were $n = 37$, $n = 59$ and, you guessed it, $n = 67$.

QUIZ QUESTION: Remember Kaprekar numbers from chapter 45? Calculate the powers of 67 up to 67^4. Is 67 a Kaprekar number for any of these powers? Bonus points if you do the calculations entirely by hand! (Okay I realise there isn't actually any points system at play here, so bonus points are particularly meaningless ... but ... um ... how about – 'I bet you don't have the guts to do it by hand!') ANSWER AT THE BACK OF THE BOOK

NAVEL GAZING

If your navel is average it should contain around 67 different types of bacteria. But the colonies of navel-inhabiting bacteria are so diverse if a friend measured theirs only a handful would be common to both of you.

On the topic of bacteria, if you were to wear a pair of jeans for two weeks without washing, in that fortnight you will have grown a colony of 1000 types of bacteria on the front, 1500 to 2500 on the back, and 10,000 on the crotch.

And while we're at it, you carry about 100,000,000,000,000 (one hundred trillion) microorganisms in your intestines. This is 10 times greater than the number of cells in your body. The vast bulk of living things inside you ... aren't you.

Next time you look at your bellybutton and meditate on what's in there, you're actually engaging in what's known as *omphaloskepsis* (navel gazing).

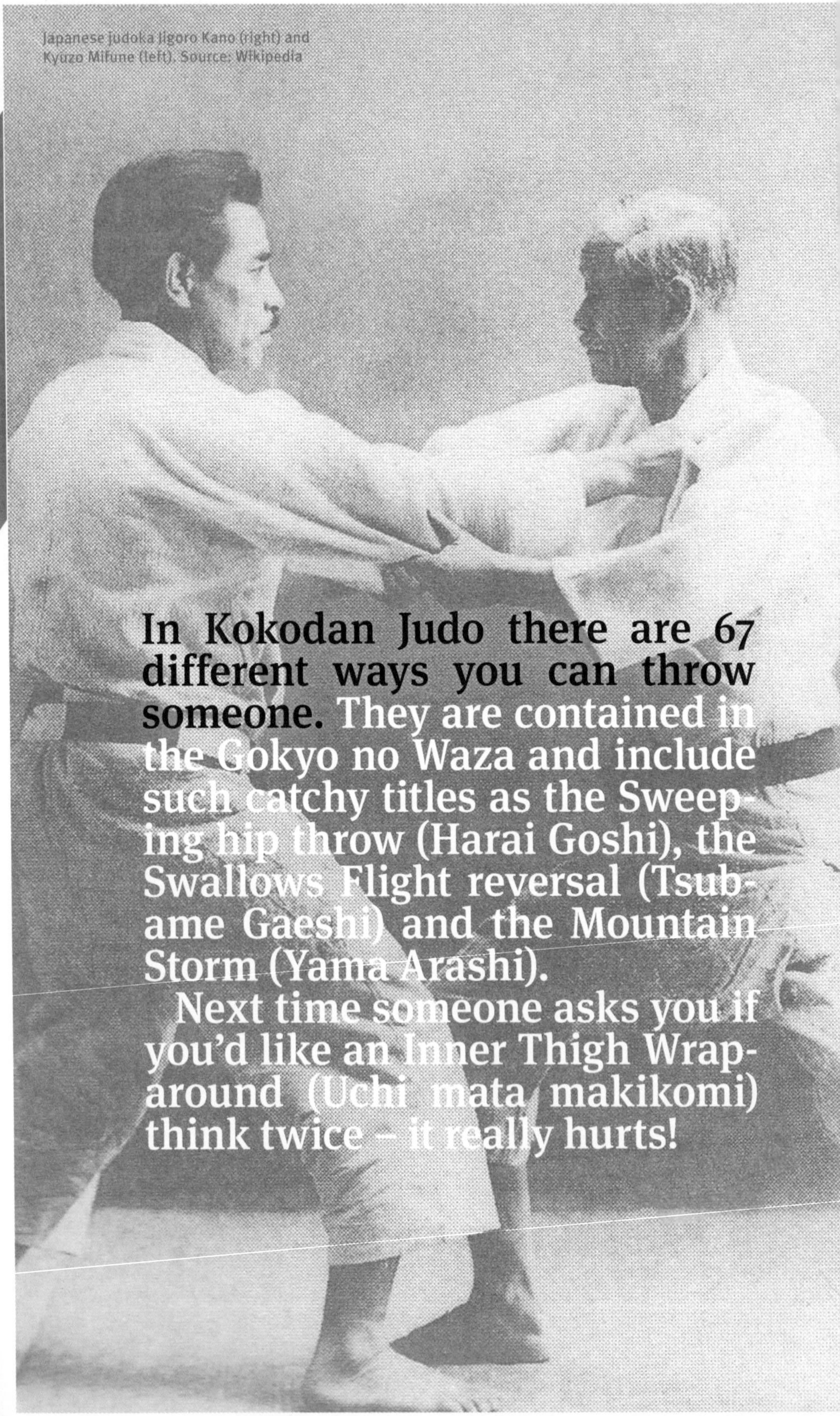

Japanese judoka Jigoro Kano (right) and Kyuzo Mifune (left). Source: Wikipedia

In Kokodan Judo there are 67 different ways you can throw someone. They are contained in the Gokyo no Waza and include such catchy titles as the Sweeping hip throw (Harai Goshi), the Swallows Flight reversal (Tsubame Gaeshi) and the Mountain Storm (Yama Arashi).

Next time someone asks you if you'd like an Inner Thigh Wrap-around (Uchi mata makikomi) think twice – it really hurts!

89

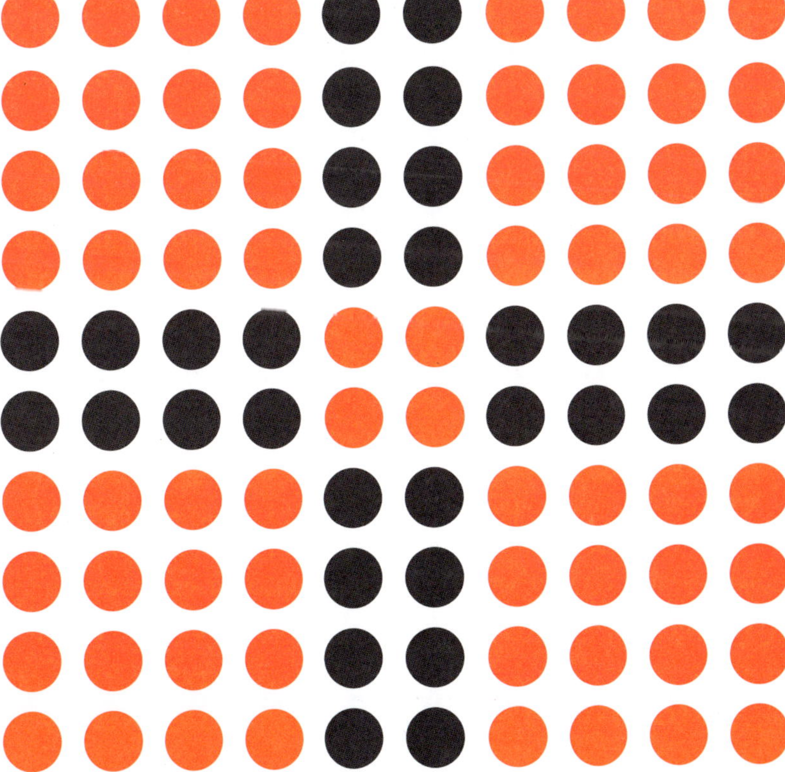

THE YEAR THAT SHOOK THE WORLD

In his book *1968: The Year that Shook the World* Mark Kurlansky argues that the congruence of the Prague Spring, the North Vietnamese's Tet Offensive, the Paris Student Riots and the assassinations of Bobby Kennedy and Martin Luther King (among other things) made '68 a particularly volatile year.

THE QUICK GREYHOUND JUMPED ...

The world record for the highest jump cleared by a dog is 172.7 centimetres (68 inches) achieved by a greyhound named Cinderella May a Holly Grey – who must also have made at least the semi-finals of weirdest dog name ever.

THE FAMOUS FOUR 4S RETURN

Back in chapter 4 we met the famous four 4s problem.

It turns out you can write the numbers 1 through 100 each just using four 4s and mathematical operations. Like I said, some of them get a bit hairy, but others aren't too bad.

When it comes to the 60s, we get some cute ones – check that you can follow them.

$60 = 4 \times 4 + 44$
$63 = (4^4 - 4) \div 4$
$64 = 4 \times (4! - 4 - 4)$
and $66 = 4 \times 4 \times 4 + \sqrt{4}$.

In particular I quite like these two renderings of 68 using four 4s:

$68 = 4 \times 4 \times 4 + 4$ and $68 = 4^4 \div 4 + 4$

CROCODILE TEETH

Humans have 20 'baby teeth' and 32 permanent teeth: 16 each in our 'maxilla' (upper jaw) and 'mandible' (lower jaw).

But crocodiles have between 64 and 68 teeth (some may have up to 80). They regularly lose them but they grow back straightaway and a croc can replace them up to 50 times in their lives.

Not only that, their eyes have a 135-degree field of view but overlap at the front by 22 degrees giving good bifocal vision.

On the topic of teeth – and lots of them – in 2014, dental surgeons removed 232 small 'tooth-like' structures from the mouth of Ashik Gavai at a hospital in Mumbai, India. Gavai suffered from a rare condition called a Complex Odontoma.

WALL STREET

While by its 200th birthday in the early 1990s it had become the international home of charcoal suits with braces, gallons of hair gel per head and some of the worst ties and vests ever seen, the world's most famous stock exchange began in 1792 underneath a tree outside number 68 Wall Street, New York, New York.

68 YEARS

On 7 May 1980, Paul Geidel walked free from the Fishkill Correctional Facility in Beacon, New York, having served time for murder. Hey, so what – that happens every day somewhere around the world. The difference here was that Geidel had entered prison on 5 September 1911, making his stay 68 years, 8 months and 2 days. That's a lot of bad mashed potato and stew.

GET HAPPY

Back in chapter 58 we played a game where we added up the squares of the digits in a number to get a new number. So:

> 58 became $5^2 + 8^2 = 89$,
> 89 became $8^2 + 9^2 = 145$, which became $1^2 + 4^2 + 5^2 = 42$,
> then 20, 4, 16 and 37 which went back to 58.

Let's try the game starting with 68.

> 68 becomes $6^2 + 8^2 = 100$
> and 100 becomes $1^2 + 0^2 + 0^2 = 1$.

And continuing to play the game just keeps giving us 1.

These are the only two things that can happen when we play this game starting with any number.

It can get caught in the loop 58, 89, 145, 42, 20, 4, 16, 37, 58, 89 ...

Or it can descend to 1.

For example, check that starting at 35 gives us the sequence 35, 34, 25, 29, 85, 89, 145, 42 ... and gets caught in the loop.

But 49, like 68, descends to 1: 49, 97, 130, 10, 1 ...

Numbers like 49 and 68 that descend to 1 are called 'happy' numbers.

QUIZ QUESTION: One and 100 are both happy numbers. Find the 18 other happy numbers from 1 to 100. ANSWER AT THE BACK OF THE BOOK

In a gloriously geeky Dr Who moment, the 10th Doctor saved the cargo ship SS *Pentallian* from crashing into the Torajii sun by deducing that the numbers 313, 331, 367 and 379 were both happy numbers *and* prime.

69

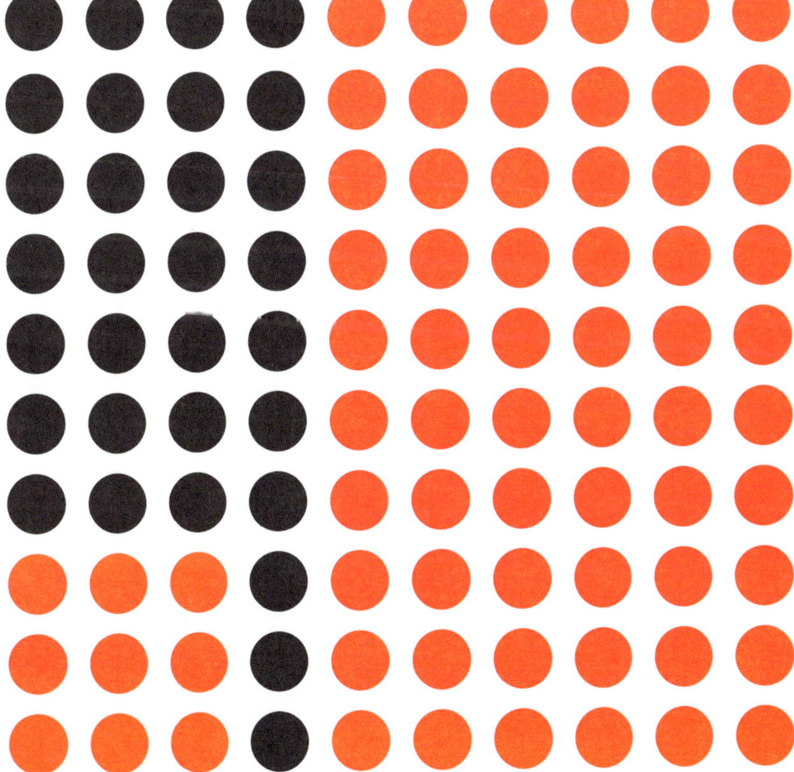

Sixty-nine is the 17th lucky number.

ROMAN NUMERALS

Derrick Neiderman is a famous American number lover, puzzle setter, cryptic crossword writer (I'm loving this guy) and a former national squash champion (okay, he lost me a bit there but hey I'm probably just jealous) and also calls himself the NumberFreak.

His excellent book of the same name showed me a few things I didn't know. Including this ...

If you write 69 out in its alphanumeric coding, that is by giving each letter its place in the alphabet, you get SIXTY NINE – 19 9 24 20 25 14 9 14 5.

Adding this up would give you $19 + 9 + ... + 5 = 139$. This doesn't have any relationship to 69 – I mean, it's $2 \times 69 + 1$ but, really, we're grasping at straws there.

However, do the same for the Roman numeral version of 69 and you get 69 = LXIX = 12 24 9 24 and guess what ... $12 + 24 + 9 + 24 = 69$.

Nice one, Derrick.

He goes onto challenge us to find the other example where this works.

QUIZ QUESTION: Apart from 69, which other number's Roman numeral depiction adds to itself alphanumerically? HINT: The number is not much less than 69. ANSWER AT THE BACK OF THE BOOK

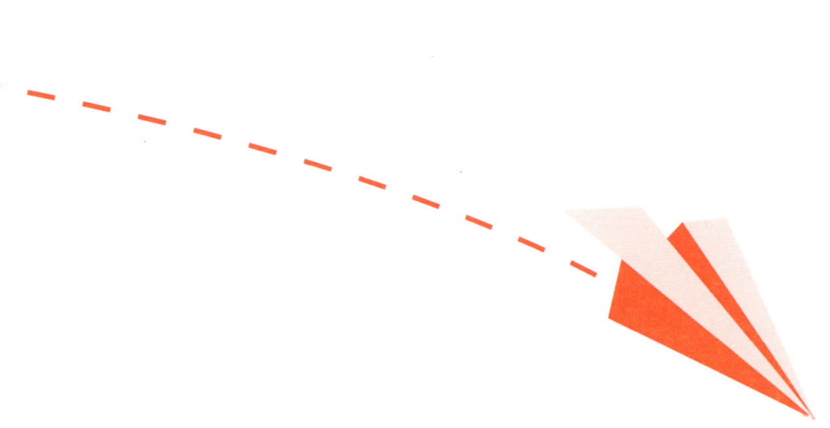

In the 80s cult film, *Bill and Ted's Excellent Adventure* (1989), the two stars confront their future selves. To test their true identity, Ted asks, 'Okay wait. If you guys are really us, what number are we thinking of?' The future Bill and Ted correctly reply, '69, dudes'. A quadruple air guitar solo follows. Who knew Keanu Reeves studied Roman numerals?

UNDERGROUND NUMBERS

The longest time survived trapped underground is an incredible 69 days by 'The 33 of San José', (32 Chilean and 1 Bolivian miners), who were trapped 688 metres (2257 feet) below the surface after the collapse of the San José copper-gold mine in Chile on 5 August 2010. As the world looked on, all 33 men made it safely back to the surface via a rescue capsule.

QUIZ QUESTION: Calculate 69^2 and 69^3 (by hand – it's good for the soul!) What unique property do you notice about the result?

ANSWER AT THE BACK OF THE BOOK

FLYING HIGH ... WELL, LONG

The furthest flight by a paper aeroplane is 69.14 metres (226 feet 10 inches), achieved by Joe Ayoob and aircraft designer John M. Collins (both from the US), at McClellan Air Force Base in California on 26 February 2012.

The fact that one guy designed the plane and another guy threw it gives an insight into exactly how serious some paper plane throwers take their craft.

Another giveaway is that John Collins has a website (paperairplaneguy.com) which has a video of Joe's record-breaking throw and the rapturous crowd reaction.

The plane was constructed from a single sheet of uncut 100 gsm A4 paper.

Another world-class paper plane thrower is Australian Dylan Parker who I've been lucky enough to meet. Perhaps reading this paragraph will inspire him to chase the world record? Parker v Collins / Ayoob ... bring it on!

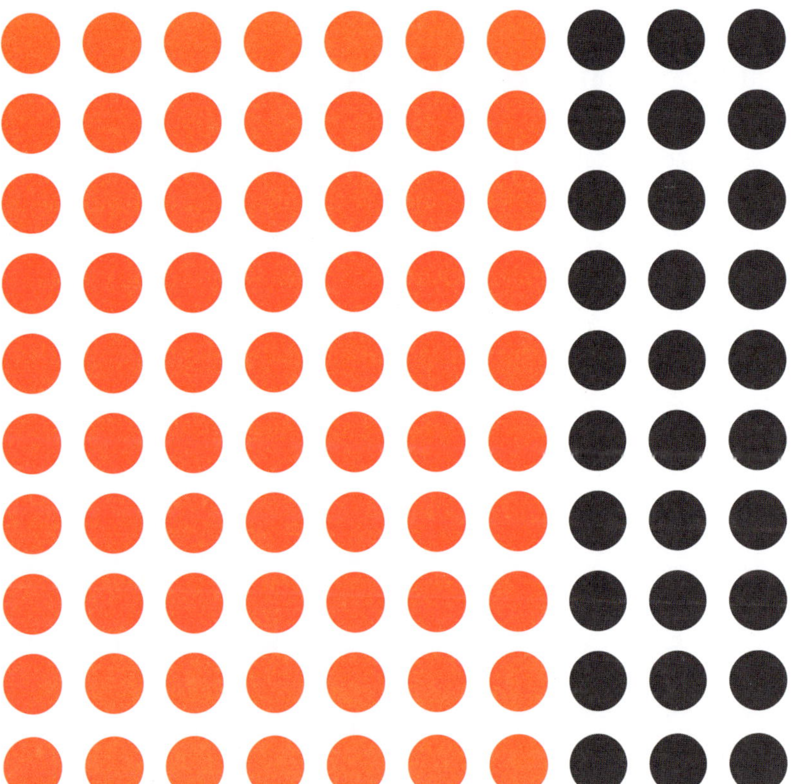

70

THE CANNONBALL PROBLEM

Sir Walter Raleigh was one busy man. A favourite of Queen Elizabeth, he was a soldier, explorer, sailor of the high seas, and was the person who introduced tobacco to Britain (which is the subject of an hilarious sketch by Bob Newhart that you really should track down one day). Just before you think, wow, how cool would it have been to be him, I should tell you that he was beheaded in 1618.

Sir Walter posed a question in mathematics, known as the cannonball problem. When a bunch of cannonballs are packed in a square pyramidal stack, how do you work out the total number of cannonballs?

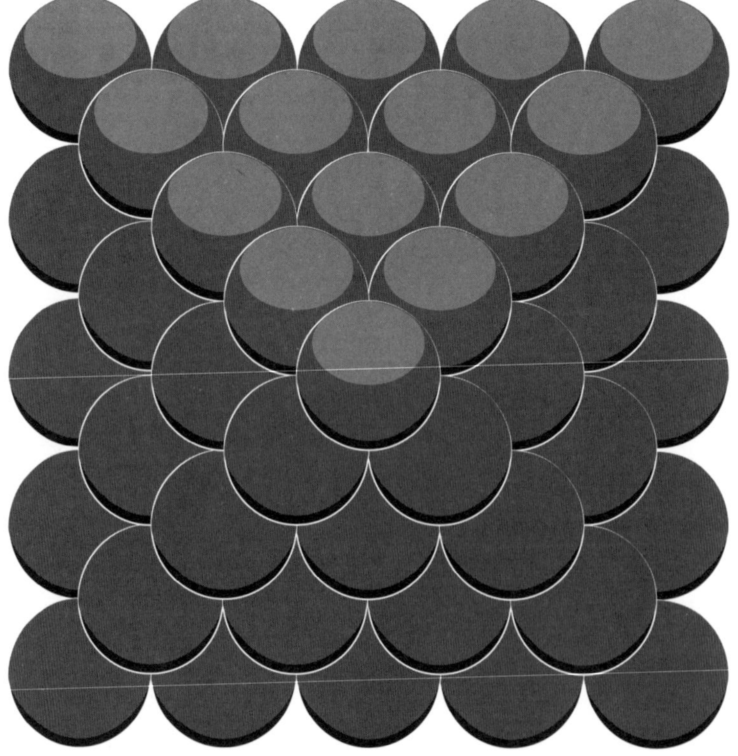

Happily, there's a formula for square pyramidal numbers that gives us an answer. If there are k layers of cannonballs, there are $\frac{1}{6}k \times (k + 1) \times (2k + 1)$ cannonballs.

If you put $k = 1$, $k = 2$, $k = 3$ into the formula, you can follow this for a few layers.

So if a captain saw a stack of cannonballs 6 layers high, they could instantly calculate that they had:

$$\frac{1}{6} \times 6 \times (6 + 1) \times (2 \times 6 + 1) = \frac{1}{6} \times 6 \times 7 \times 13 = 91 \text{ cannonballs.}$$

QUIZ QUESTION: There is only one square pyramidal stack of cannonballs with a number of cannonballs that is itself a square. That pyramid has 24 layers. How many cannonballs are in the pyramid? ANSWER AT THE BACK OF THE BOOK

WEIRDO

Seventy is a 'weird' number, but not because it's kinky or anything. A weird number is one that is abundant but isn't equal to the sum of any of its divisors (that is, it's abundant but not semiperfect – see chapter 54). Weird numbers are very rare, and 70 is the only one below 100. For example, 12 is abundant because $1 + 2 + 3 + 4 + 6 = 16$, which is greater than 12. But $2 + 4 + 6 = 12$, so 12 isn't weird. See if you can convince yourself that 70 is.

Oh, and the longest game of Monopoly on record lasted 70 days. Now, *that's* weird!

The average Komodo dragon weighs 70 kilograms. So if you're looking at grabbing a strapless, red, off-the-shoulder number for a Komodo dragon mate of yours, a size 14 is probably a safe, loose-fitting bet.

FOR WHOM THE PELL TOLLS

Seventy is a Pell number (see chapter 29). The first few Pell numbers are 1, 2, 5, 12, 29, 70, 169, 408 … and they arise from trying to 'guess' or approximate $\sqrt{2}$.

There doesn't seem to be any obvious rule as to how to get from one Pell number to the next. We don't just double them or add two numbers to get the next like the Fibonacci sequence. But notice that if you double a Pell number and add the one before it you get the next number.

Huh? Well, to get the Pell number after 29, double 29 and add the number before it: $29 \times 2 + 12 = 70$

In fancy-pants mathematics talk we say the Pell numbers satisfy what's known as the 'recurrence relation':

$$P_0 = 0,\ P_1 = 1 \text{ and } P_n = 2 \times P_{(n-1)} + P_{(n-2)}$$

Aaaaaarrgghhh! … Don't worry – it's not as scary as it looks.

We start with $P_0 = 0$ and $P_1 = 1$ and putting $n = 2$ into the rule, we get: $P_2 = 2 \times P_1 + P_0 = 2 \times 1 + 0 = 2$

$n = 3$ gives us $P_3 = 2 \times P_2 + P_1 = 2 \times 2 + 1 = 5$

$n = 4$ gives $P_4 = 2 \times P_3 + P_2 = 2 \times 5 + 2 = 12$ … and so on.

Convince yourself that the next two Pell numbers are 29 and 70.

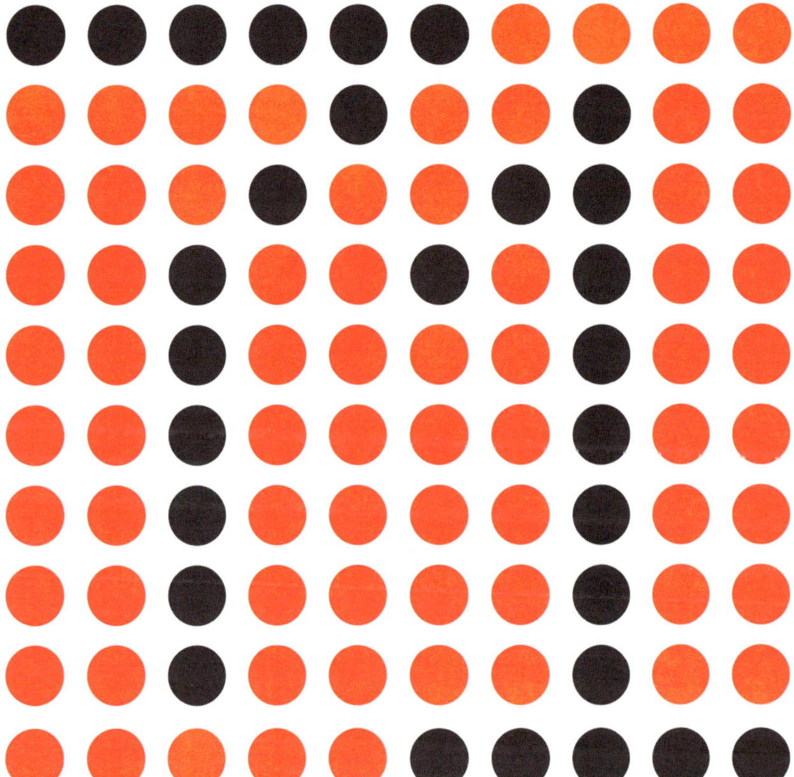

71

FERMAT-CATALAN CONJECTURE

So, $2^7 + 17^3 = 71^2$.

You can work this out by yourself. If you're really keen, you don't even need a calculator ... a good old pen and paper will do. Go on, you know you want to!

You get:

$128 + 4913 = 5041$

... which is certainly correct.

And you can check that 2, 17 and 71 don't have any common factors – this is obvious because they are all prime. From chapter 16 you might remember this means that 2, 17 and 71 are relatively prime or 'coprime'.

You can also do something else pretty neat, if you can remember fractions from the early days of high school. Look at the powers in this equation: they are 7, 3 and 2.

Now invert them (don't freak out – that just means make them a fraction with 1 on the top) and add them up:

$\frac{1}{7} + \frac{1}{3} + \frac{1}{2} = \frac{6}{42} + \frac{14}{42} + \frac{21}{42}$ (remember we make them all fractions over 42 so we can add them up) = $\frac{41}{42}$ which is less than one.

So, to put it a bit more fancily ...

$2^7 + 17^3 = 71^2$ is an equation of the form $x^p + y^q = z^r$ where x, y and z are positive, coprime integers, and p, q and r are positive integers satisfying $\frac{1}{p} + \frac{1}{q} + \frac{1}{r}$ less than or equal to 1.

The Fermat-Catalan conjecture is named after two famous mathematicians that we've heard of in this book – which is pretty impressive given that they never met (they lived hundreds of years apart!) The conjecture says that there are only a finite number of equations that tick all the boxes that this one does.

As of 2008 only 10 had been found:

$1^6 + 2^3 = 3^2$ (Catalan)

$2^5 + 7^2 = 3^4$

$7^3 + 13^2 = 2^9$

$2^7 + 17^3 = 71^2$

$3^5 + 11^4 = 122^2$

$17^7 + 76{,}271^3 = 21{,}063{,}928^2$

$1414^3 + 2{,}213{,}459^2 = 65^7$

$9262^3 + 15{,}312{,}283^2 = 113^7$

$43^8 + 96{,}222^3 = 30{,}042{,}907^2$

$33^8 + 1{,}549{,}034^2 = 15{,}613^3$

Strictly speaking, we could replace the 1^6 with 1^7, 1^8 and so on, and get an infinite number of equations but ... we don't, okay.

HEY, BROCARD

We've met factorials before (see chapter 24).

So $4! = 4 \times 3 \times 2 \times 1 = 24$.

And $5^2 = 5 \times 5 = 25$.

So $4! + 1 = 5^2$.

A fancier way to say this is that the numbers $m = 5$ and $n = 4$ satisfy the equation $n! + 1 = m^2$. Finding whole numbers, or integers, that satisfy this equation is called the Brocard problem.

We also know that $m = 11$, $n = 5$, and one other pair of numbers satisfy the Brocard problem.

A brilliant and eccentric mathematician named Paul Erdös thought that there are only three solutions to the Brocard problem. The latest info comes from two dudes, Bruce Berndt and William Galway, who have checked for n up to $n = 1{,}000{,}000{,}000$ and found no other solutions – so Erdös may well be on a winner here.

QUIZ QUESTION: For the third solution to the Brocard problem, n is less than 10. Find n and m. ANSWER AT THE BACK OF THE BOOK

Seventy-one per cent of the Earth's surface is covered in water – but considerably less than 1% of all the water on Earth is freshwater that we can actually use.

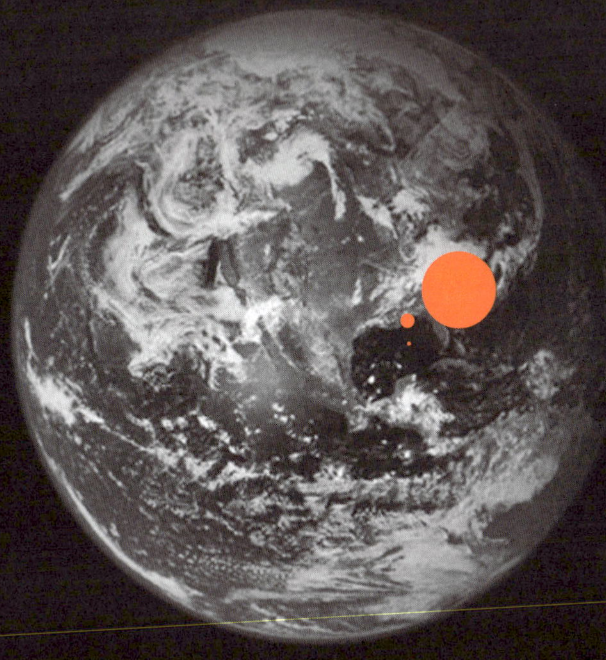

A sphere approximating all of the Earth's water (oceans, ice caps, lakes, and rivers, groundwater, atmospheric water, and even the water in 'living things' – that includes you, buddy). Its volume would be about 1,386,000,000 km^3.

A sphere approximating all liquid freshwater (groundwater, lakes, swamps, rivers). The volume amounts to about 10,633,450 km^3 – although 99% of that is groundwater, deposited beneath the Earth's surface and therefore not-at-all easily accessible to humans.

A sphere approximating all freshwater in lakes and rivers – the kind we can access relatively easily. The volume would be about 93,113 km^3.

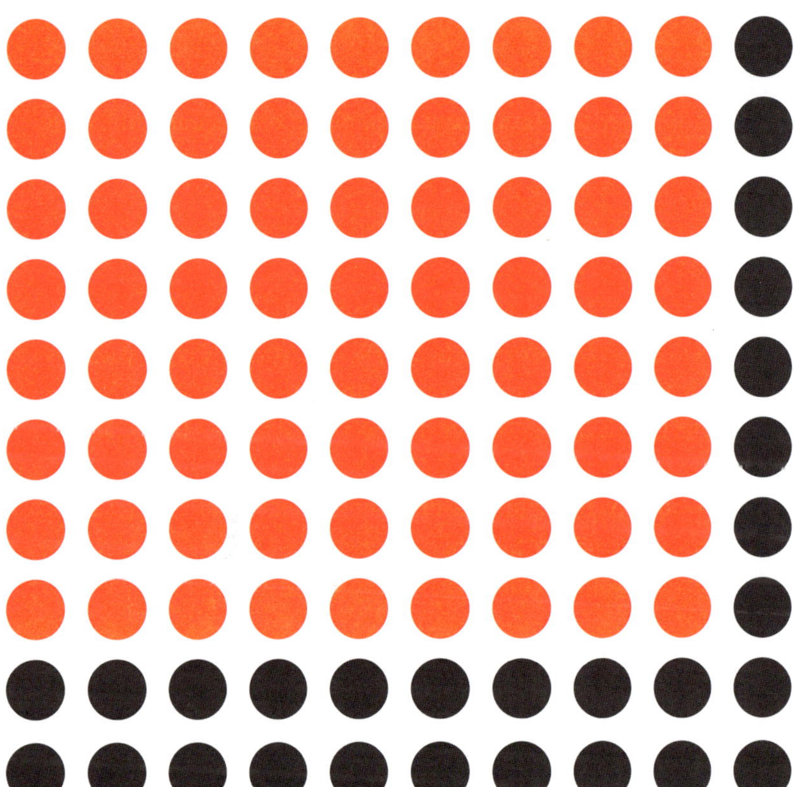

72

It is often quoted that you use 72 muscles in your mouth when you talk. If that's true, coming out with, 'Um ... yeah ... um ... ah' really seems a bit of a waste, doesn't it?

$72 = 2^3 \times 3^2$

Further, 72 to the power of five (72^5), or 1,934,917,632 as you may know it, equals $19^5 + 43^5 + 46^5 + 47^5 + 67^5$.

In fact, 72^5 is the smallest fifth power equal to the sum of five fifth powers.

A couple of the solutions to writing 72 using only four 4s (see chapters 4 and 68) are particularly cute.

Convince yourself that:

$$72 = (4! \times 4!) \div (4 + 4) = 44 + 4! + 4 = 4 \times (4 \times 4 + \sqrt{4})$$

A BEAUTIFUL FILM

There are 72 hours in three days, as beautifully encapsulated in the title of the Russell Crowe film, *72 hours – The Next Three Days*.

However, this is certainly not Russell Crowe's only filmic fusion of art and mathematics. For mine, his superb portrayal of the brilliant but troubled American John Nash in *A Beautiful Mind* was scandalously overlooked at the 2001 Academy Awards – no offence intended to the very talented Denzel Washington and his performance in *Training Day*.

QUIZ QUESTION: Seventy-two can be written as the sum of 4 consecutive primes. It can also be written as the sum of 6 consecutive primes. Find both ways to write 72 as the sum of consecutive primes. ANSWER AT THE BACK OF THE BOOK

INTERIOR ANGLES

By dividing a regular pentagon into three triangles (and remembering that the angles in a triangle add up to 180 degrees) we can work out the following:

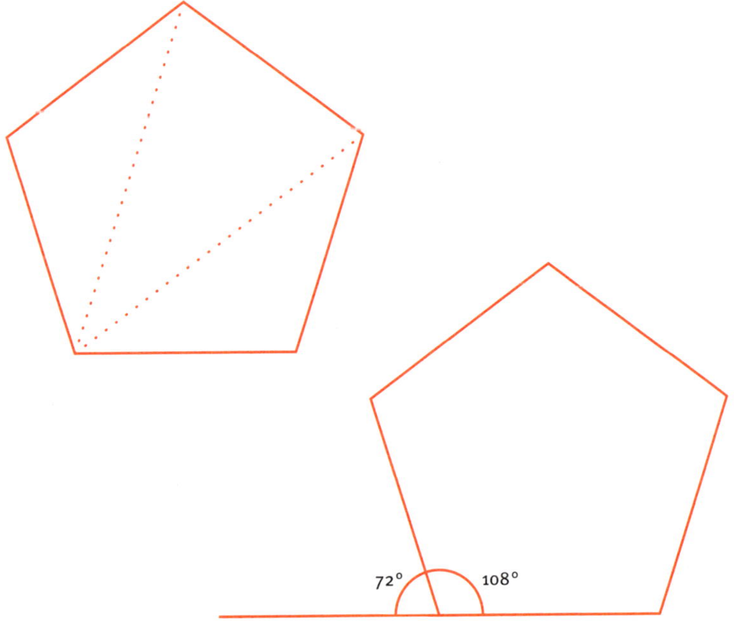

The angles in the three triangles add up to $3 \times 180 = 540$ degrees.

But these angles also cover exactly the 5 angles inside the pentagon. Those 5 angles are all equal so must each be $\frac{540}{5} = 108$ degrees.

So the 'interior' angles of a regular pentagon are 108 degrees and we also know the angle in a straight line is 180 degrees.

So the 'exterior' angles of a regular pentagon are 72 degrees.

You can follow the same method to work out the exterior and internal angles of equilateral triangles and squares (obvious) and regular hexagons, heptagons, octagons and so on.

While Angkor Wat in Cambodia is the probably the most famous and spectacular example, there are actually 72 temples in the Angkor region.

These architectural masterpieces were built between 900 and 1200 AD and are visited by over 2 million tourists a year. So if you pop along to Angkor, please look after it – it's the only one we have.

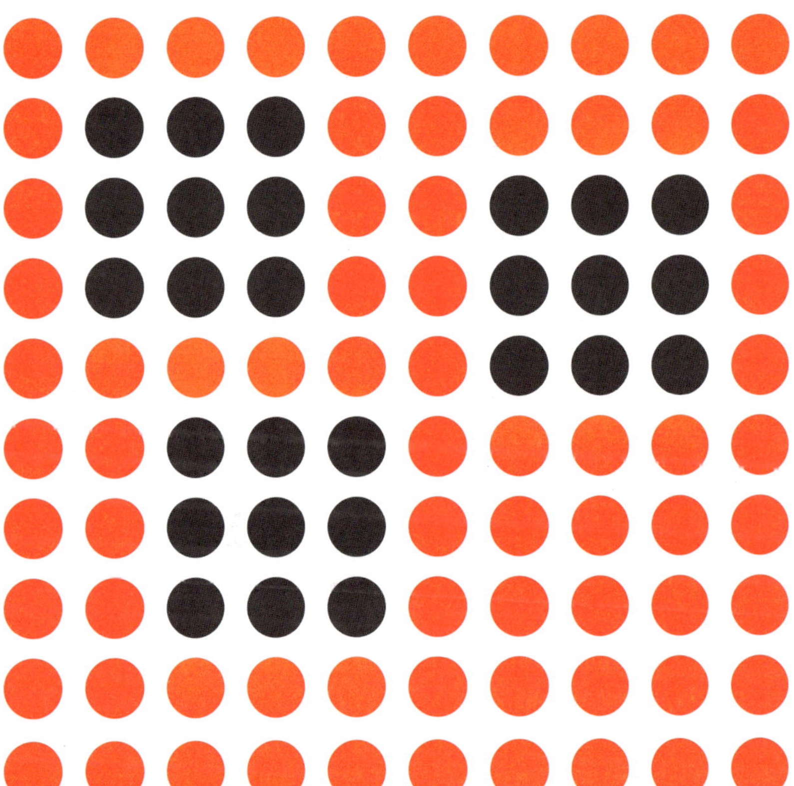

73

THE BIG BANG THEORY

I've spent a lot of time in this book giving you my opinions about what's hot in the world of numbers. But other people have their own views. In the international TV sensation *The Big Bang Theory* 'Alien Parasite Hypothesis' episode, lead geek Dr Sheldon Cooper shares this:

SHELDON: What is the best number? By the way, there's only one correct answer.
RAJ: 5,318,008?
SHELDON: Wrong! The best number is 73. [Short silence] You're probably wondering why.
LEONARD & HOWARD: No no, we're good.
SHELDON: Seventy-three is the 21st prime number, its mirror 37 is the 12th and its mirror 21 is the product of multiplying, hang on to your hats, 7 and 3. Did I lie?
LEONARD: We did it! Seventy-three is the ... Chuck Norris of numbers!
SHELDON: Chuck Norris wishes! In binary, 73 is a palindrome, 1001001, which backwards is 1001001, exactly the same. All Chuck Norris gets you backwards is Sirron Kcuhc!
RAJ: Just for the record, when you enter 5,318,008 in a calculator, upside down it spells BOOBIES!

Geek gold.

A beautiful, hidden piece of arithmetic in this exchange is that 'Alien Parasite Hypothesis' was the 73rd episode of *The Big Bang Theory*. Further, Jim Parsons, the actor who plays Sheldon Cooper, was born in 1973 so when the episode featuring the homage to 73 / 37 went to air, on 9 December 2010, he was 37 years old.

Seventy-three is the smallest whole number with 12 letters in its name.

SIERPIŃSKI NUMBERS

No matter what value of n you choose, the expression:

$78{,}557 \times 2^n + 1$ is never prime. It will always have 3, 5, 7, 13, 19, 37 or 73 as a factor. For example:

$$78{,}557 \times 2^1 + 1 = 157{,}115 = 31{,}423 \times 5$$
$$78{,}557 \times 2^2 + 1 = 314{,}229 = 104{,}743 \times 3$$
$$78{,}557 \times 2^{11} + 1 = 160{,}884{,}737 = 12{,}375{,}749 \times 13$$

... and so on for any value of n.

When you think about how little we know about prime numbers it's really quite amazing that no matter what n you put into the equation, you will never get a prime answer.

Now, 78,557 isn't the only number k for which $k \times 2^n + 1$ is never prime, in fact, the Polish mathematician Wacław Sierpiński proved that there are an infinite number of odd numbers k. But other mathematicians strongly believe that 78,557 is the smallest such Sierpiński number.

In 2002 an organised distributed computing attack was launched to examine the last 17 cases of numbers less than 78,557 that were not known to generate a prime. It is run through the website seventeenorbust.com and has now reduced the list of possible candidates to single figures.

QUIZ QUESTION: Show that $78{,}557 \times 2^3 + 1$ is divisible by 73. You can use a calculator if you reeeaaaally want, but if, like me, you've got a soft spot for long division, here's your chance to go crazy.

ANSWER AT THE BACK OF THE BOOK

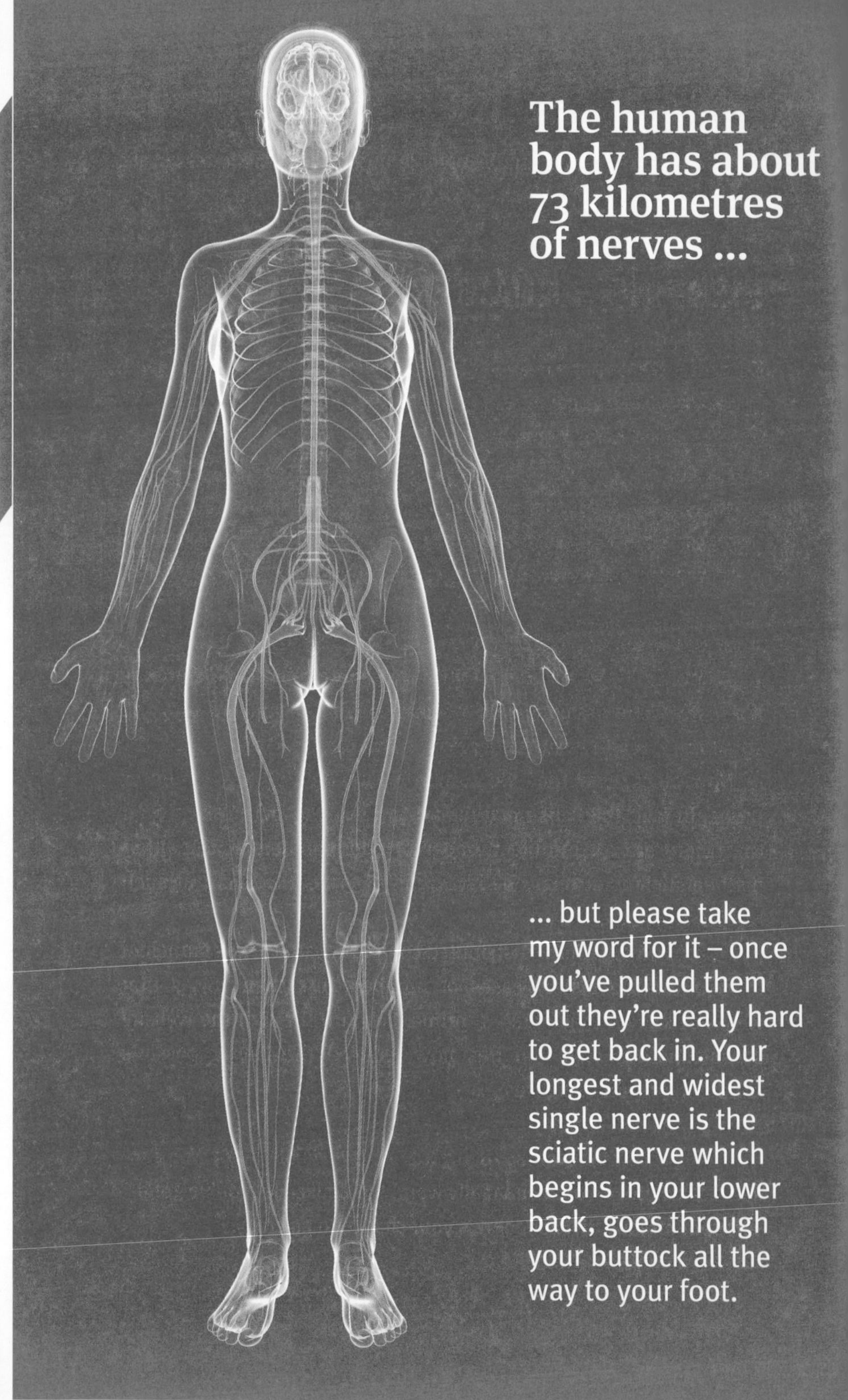
The human body has about 73 kilometres of nerves ...
... but please take my word for it – once you've pulled them out they're really hard to get back in. Your longest and widest single nerve is the sciatic nerve which begins in your lower back, goes through your buttock all the way to your foot.

KEPLER SPHERE PACKING

If you stack a group of tennis balls – or cannonballs if you're a 17th century pirate (in which case, you might also like to check out chapter 70) – in this manner, you're actually stacking them as efficiently as possible.

Although Johannes Kepler (the same guy who worked out that the planets travel elliptically around the Sun) suggested this in 1611, it took until 1998 and the advent of supercomputers to prove him correct.

This 'Kepler packing' of spheres has an efficiency or density of $\frac{\pi}{3\sqrt{2}}$ or pretty much 74%.

But if we allow ourselves to move away from spheres there are other objects that pack together even more efficiently.

In 2004, a team of mathematicians led by Aleksander Donev from Princeton in the US, had produced some potentially important results

KHMER

is the official language of Cambodia. It has the most letters of any alphabet – 74. (The shortest, for the record, is the Rotokas language of Papua New Guinea, with only 12 letters.)

G'DAY, MATE

To all the international readers of this book let me do my bit for Australian tourism and say that the incredibly beautiful Whitsundays (off the coast of Queensland and covering the Great Barrier Reef) contain 74 islands.

THE MOST TATTOOED

senior citizen is Tom Leppard from the UK, aged 74, who has 99.9% of his body covered. Tom, who resides on the Isle of Skye, opted for a leopard-skin design, with all the skin between the dark spots tattooed saffron yellow ... and no, Tom, I don't want to know where the other 0.1% is!

by comparing randomly packed spheres with ellipsoids.

See, when we lay spheres out one by one in the perfect Kepler packing we get that 74% efficiency figure, but if we drop spheres randomly into a big container they most likely won't fall into a Kepler packing. They'll jam up against each other with more like 64% efficiency and on average each sphere will only touch about 6 others before it gets jammed.

When we drop 'ellipsoids' into a container randomly they fill at more like 68 to 74% – the sort of efficiency spheres can only get by being deliberately placed.

What, you may well ask, is an ellipsoid? It looks a bit like a sphere that has been squashed. But Donev gives an example that might be easier to understand ... M&Ms or Smarties are both ellipsoids.

Donev and his team 'performed experiments with M&Ms' and found that a typical M&M touches 10 others before it gets jammed and this extra degree of freedom is why they can be more tightly packed. M&Ms pack at around 68%, other ellipsoids up to 74%.

Some of the younger readers of this book may not have been around in a time long past – not quite as far back as when dinosaurs roamed the Earth – when music came on something called 'compact discs'.

We gave compact discs the trendy abbreviation CD (because we were pretty swag!) Anyway, the original CD was 12 cm in diameter and held its information in 16-bit format. The result was it could hold 74 minutes of music.

One popular theory for why CDs were made with 74 minutes capacity is that the Chairman of Sony, Akio Morita, wanted CDs to be able to hold even the slowest possible version of Beethoven's 9th Symphony – because it was his wife's favourite piece of music. This claim has never been proven but if it's true, good on you, Akio, you romantic old devil you!

Portrait of Ludwig van Beethoven in 1820. Source: Public Domain

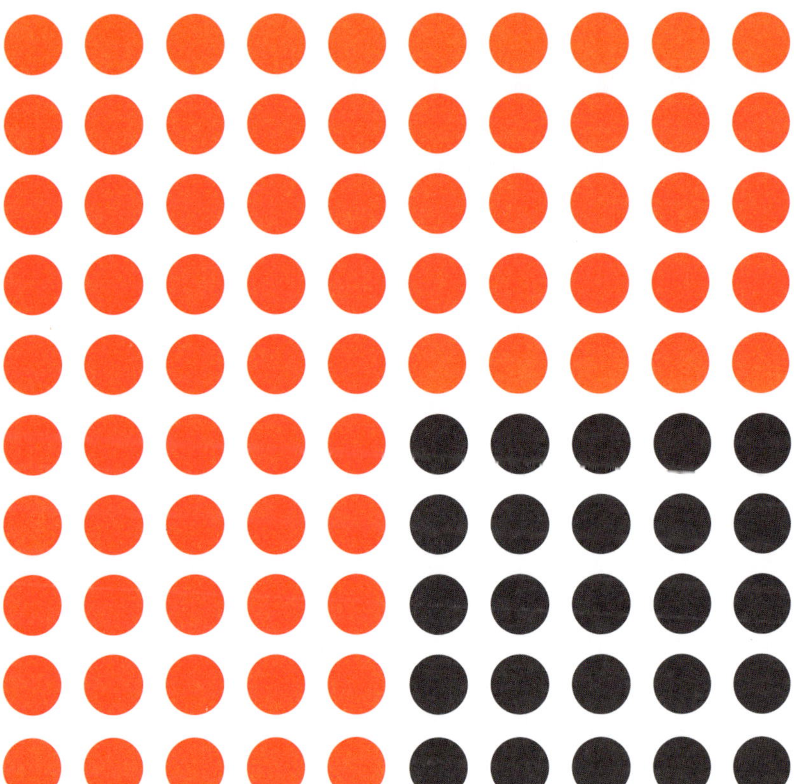

75

PENTAGONAL PYRAMIDAL NUMBER

The first five pentagonal numbers (see chapter 22) are:

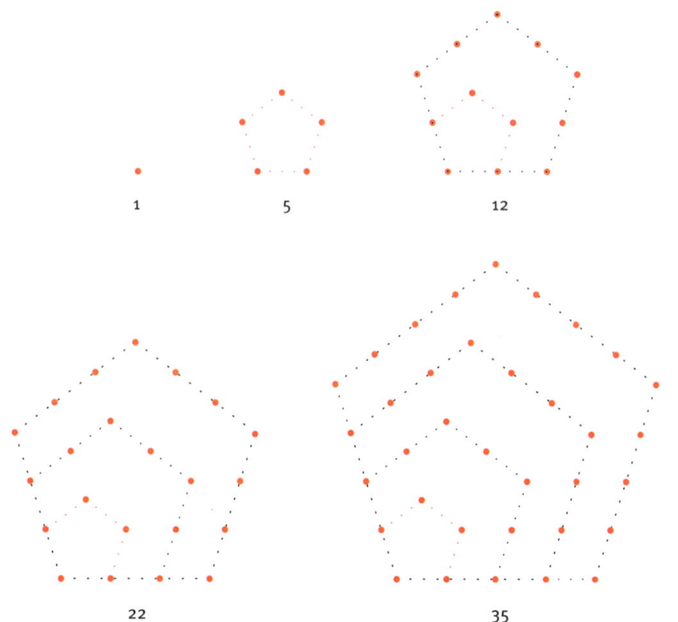

And we've seen in chapters 55 and 56 that when we add triangular or square numbers we can make 'pyramidal numbers'. So $1 + 5 + 12 + 22 + 35 = 75$, and we call 75 the fifth 'pentagonal pyramidal' number.

When I say 'we' I'll be honest, I very rarely use the phrase in conversation ... and I presume you're the same. Let's promise to drop it this weekend and turn a few heads.

QUIZ QUESTION: The pentagonal pyramidal numbers 1, 6, 18, 40, 75, 126, 196 ... are given by the formula $P(n) = \frac{n^2(n+1)}{2}$. So for $n = 3$, $P(3) = \frac{3^2(3+1)}{2} = \frac{9 \times 4}{2} = 18$. Starting with $n = 1$ then $n = 2$ and so on, generate the first 10 pentagonal pyramidal numbers.

ANSWER AT THE BACK OF THE BOOK

SELF NUMBERS

Seventy-five is the 19th lucky number which we've heard about in chapter 25; it's a Keith number (chapter 47) but it's also an example of a pretty obscure sort of number called a 'self' number.

It's easier to explain a self number by first showing a few that *aren't*.

We can write 30 as 30 = 24 + 2 + 4 that is, we wrote 30 as the sum of a number (24) and the digits of that number (2 and 4).

Similarly, 8 = 4 + 4; 46 = 41 + 4 + 1; 83 = 73 + 7 + 3; 99 = 90 + 9 + 0

But there is no number that when you add it to its digits you get 75. That's why we call 75 a self number.

QUIZ QUESTION: One is a self number and there are 12 others between 1 and 100. Find them. (And this really is only for the keen among you. If you skip this one I won't judge you poorly at all. It's not hard – just long!) HINT: It's easier to find all the non-self numbers. Start by adding 1 to the sum of its digits; then add 2 to the sum of its digits; then 3, then 4 and so on. Each time you do this you get a non-self number and you can cross that off the list of numbers 1 to 100. Whatever's left are the self numbers. ANSWER AT THE BACK OF THE BOOK

KAPOW!

French 75 is the name of a cocktail first recorded in *The Savoy Cocktail* book (1930) and is made from gin, champagne, lemon juice and sugar. It was created in 1915 at the New York Bar in Paris and was said to have such a kick that it felt like being shelled with the powerful French 75-millimetre field gun.

You would know that when an organisation or country turns 100 it celebrates its 'centenary'. So it's not surprising that for 200 years it's a *bi*centenary, what with the prefix 'bi' meaning 2 and all. But, according to a handful of websites, when we celebrate a 75th birthday, we're getting together for a *dodranscentennial*. Dodrans comes from the Latin de-quadrans which means 'a whole part minus a quarter' and hence $\frac{3}{4}$. Three-quarters of a century is 75 years, hence, they claim, dodranscentennial.

These things all depend on levels of usage, so why don't you join my campaign, starting here, to get the d-word into mainstream English?

Hey, grandpa, happy dodranscentennial!

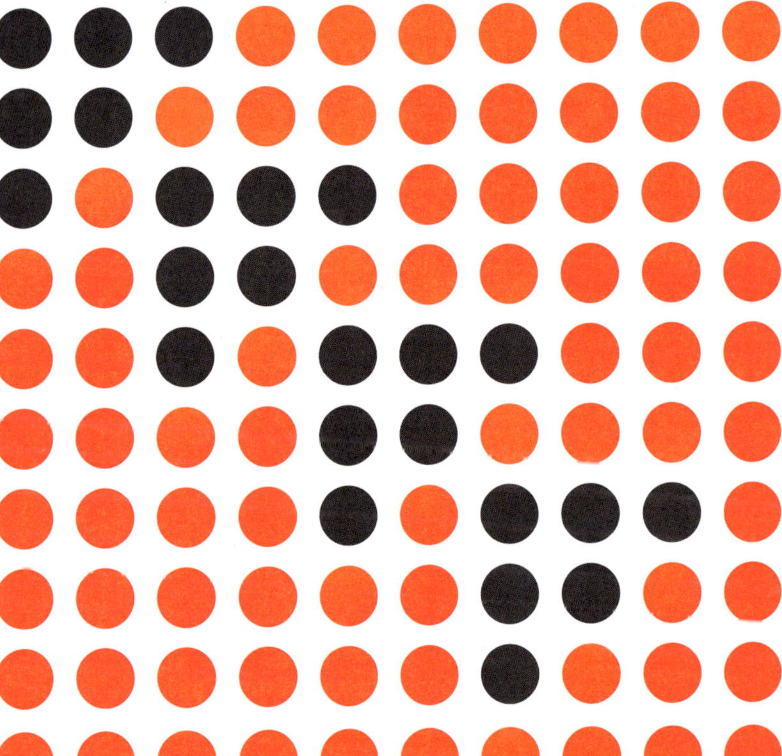

76

TELEPHONE NUMBERS

Imagine the four dots in the diagrams below are all people connected to a phone service. A dark line between two dots means those two people are talking on the phone. Obviously at any given time not everyone needs to be on the phone, but this town doesn't have call waiting so you can't be 'talking' to two people at once.

There are 10 ways this network can be running from:

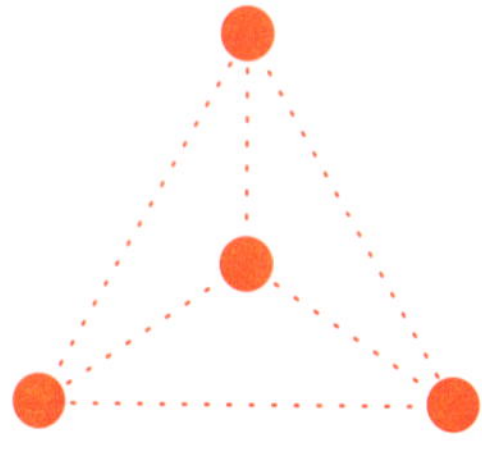

... where no-one is on the phone, through 6 single phone calls with the other two people not on the phone:

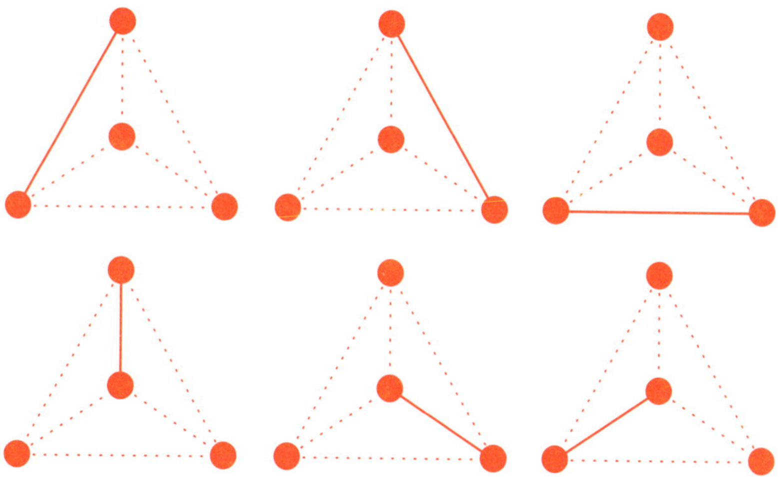

Up to 3 ways that two calls are happening meaning everyone is involved:

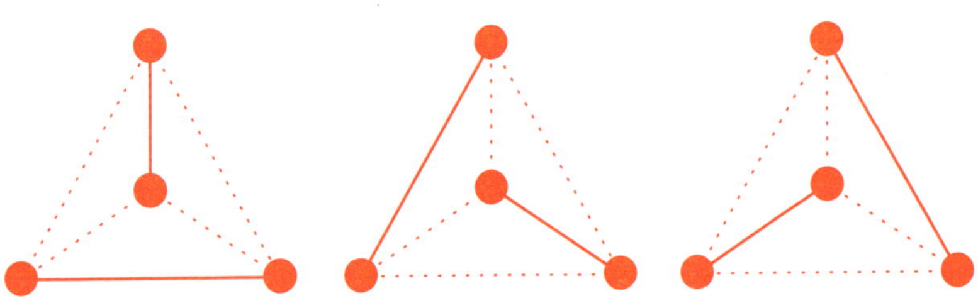

These 10 possible call patterns mean that 10 is the fourth 'telephone' number.

Starting with one poor lonely person who can't talk to anyone:

... going to two people who may be on the phone to each other or not:

... and moving up through bigger populations, the telephone numbers are 1, 2, 4, 10 , 26 and the sixth is 76.

Though it might seem we're just making up silly types of numbers here, telephone numbers actually have important applications in several areas of mathematics that involve counting of various objects.

Seventy-six is a bit of a stretch, but try and find the 26 call patterns for a 5-person grid.

Austrian composer Arnold Schoenberg suffered *triskaidekaphobia* (see chapter 13) and was convinced that he'd die at $65 = 13 \times 5$. He lived into his seventies but endured a torturous year, aged 76, after a friend pointed out that $7 + 6 = 13$. He died at a quarter to midnight 15 minutes before he was to turn 77.

HOORAY HALLEY

The English scientist Edmond Halley was born in 1656 and lived to the ripe old age of 86 which was pretty impressive for his time.

Halley's father was a wealthy soap-maker and the young Edmond loved mathematics. He invented a diving bell in which he once remained under the Thames for over 4 hours, made breakthroughs in accounting and finance, could translate Arabic into English and eventually became Britain's second-ever Astronomer Royal.

But Ed is most famous for suggesting that a series of comet sightings in 1456, 1531, 1607, and 1682 were all actually the same comet, which he predicted would return in 1758. He died before the comet's return, but when it did come back, it was christened Halley's Comet.

Halley's Comet is probably the most famous of all comets and is visible from the Earth once every 76 years. The last sighting was 9 February 1986. However, there is much debate about whether the Comet's orbit is 75 or 76 years.

Other cool-sounding comet names include Shoemaker-Levy, Pons-Winnecke, Giacobini-Zinner and Schwassman-Wachman.

I admit I'm biased on this subject, but my favourite piece of space rock is asteroid 18413 Adamspencer named by the Australian planet discovering guru Robert H. McNaught. It just chugs along between Mars and Jupiter doing not very much, but I'm extraordinarily proud to have had it named after me.

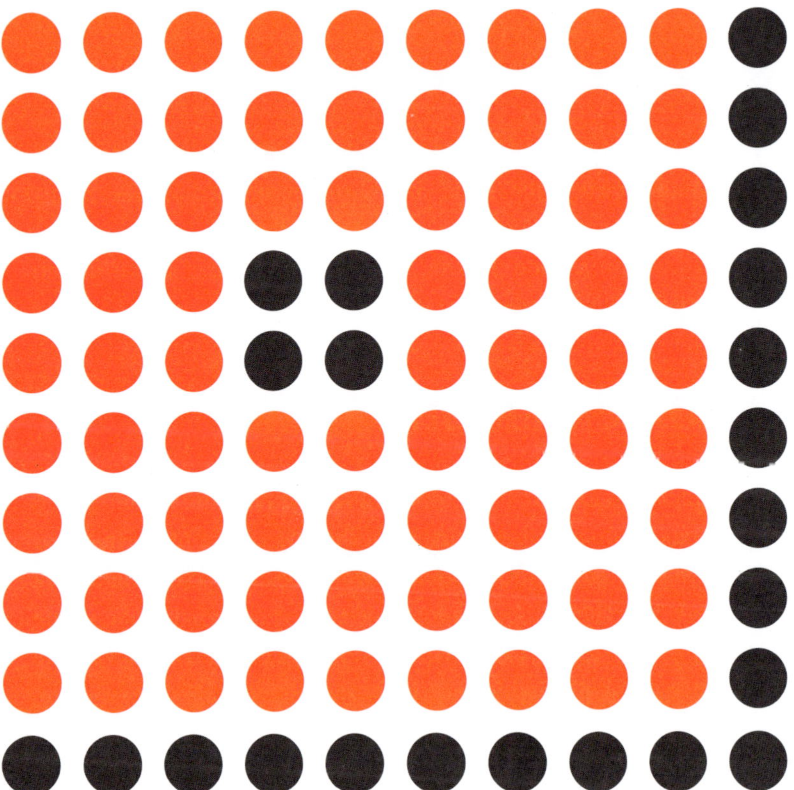

77

How's this? $7^7 + 77^7 + 777^7 + 2 = 170{,}980{,}732{,}128{,}390{,}323{,}351$ is a prime number whose digits add up to ... are you sitting down? ... 77.

The art punk band Talking Heads released their debut album titled *Talking heads: 77* in 1977.

ANOTHER MASSIVE PRIME

Remember factorials? If you don't, have a quick squiz at chapter 6 to remind yourself. Now, it turns out that 77! + 1 is a prime number. Trust me: it has over 100 digits, so it would be an absolute nightmare to work out. The numbers n for which $n! + 1$ is known to be prime begin with $n = 1, 2, 3, 11, 27, 37, 41, 73, 77$ and currently end with the massive 150,209! + 1 which is 712,355 digits long.

I know you're curious, so here you go: 77! + 1 = 145,183,092,028,285,869,634,070,784,086,308,284,983,740,379,224,208,358,846,781,574,688,061,991,349,156,420,080,065,207,861,248,000,000,000,000,000,001

SOME SEVENTY-SEVEN SNIPPETS

There are 77 partitions of 12, which isn't hard to prove but takes time and some attention to detail (see chapter 4).

While we're on the subject, convince yourself that the sum of the first 8 prime numbers is 77, remembering that 1 is *not* prime (see chapter 1).

Now that you're warmed up ...

QUIZ QUESTION: Write 77 as the sum of 3 squares, two different ways. ANSWER AT THE BACK OF THE BOOK

SJUTTIO SJU

In English, 77 is the smallest positive integer that requires 5 syllables.

But in Swedish, 77, or as they like to say, *sjuttiosju*, is a bugger to pronounce, so in World War II it was a popular password on the Sweden–Norway border. Soldiers could easily tell if the speaker was a native Swede, Norwegian or German.

PRIMETIME 77

In chapter 94 we'll meet Christian Goldbach and Goldbach's 'weak conjecture'. A consequence of this conjecture states that all odd numbers can be written as the sum of at most three primes.

For prime numbers we obviously can write them as the sum of just one prime – themselves, for example, 37 = 37.

And for every prime number we can write the next odd number as it plus 2, for example 55 = 53 + 2, 69 = 67 + 2.

But some odd numbers need all three primes allowed by Goldbach's weak conjecture when expressing them. One such prime is 77.

You can write 77 = 3 + 3 + 71 = 5 + 5 + 67 = 3 + 7 + 67 = 11 + 29 + 37 and lots of other ways.

But because 77 is not prime and it is two more than 75 and 75 is not prime you can't write it by itself nor as a prime plus two. So you need all three primes. Easy!

QUIZ QUESTION: **Seventy-seven is one of nine odd composite numbers between 1 and 100 that cannot be expressed as the sum of two primes. Find the others.** HINT: Write down all the odd numbers between 1 and 100 and eliminate the primes and all numbers equal to a prime plus two.

ANSWER AT THE BACK OF THE BOOK

In 2005 at the 77th Academy Awards, Cate Blanchett won the Oscar for Best Supporting Actress for *The Aviator*, in which she played the great Katharine Hepburn who, herself, won four Best Actress Oscars between 1934 and 1982.

SUDOKU

We saw earlier how a Sudoku needs at least 17 entries to have a unique answer.

At the other extreme, Jean-Paul Delahaye's great 2006 article in *Scientific American* gives the Sudoku below. There is no unique solution despite the 77 'givens' because the top row could be 1, 2 and the fourth row 2, 1 or vice versa.

		3	4	5	6	7	8	9
4	5	6	7	8	9	1	2	3
7	8	9	1	2	3	4	5	6
		4	3	9	8	5	6	7
8	6	5	2	7	1	3	9	4
9	3	7	6	4	5	8	1	2
3	4	1	8	6	2	9	7	5
5	7	2	9	1	4	6	3	8
6	9	8	5	3	7	2	4	1

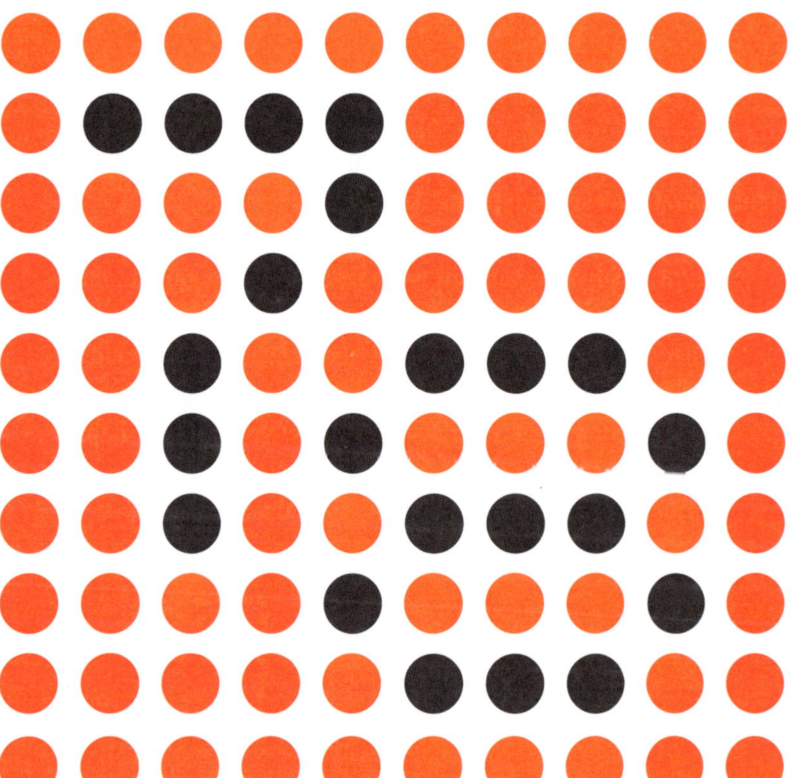

78

THE 12TH TRIANGLE

Seventy-eight is an abundant number. It's also the 12th triangular number. Let's play with triangular numbers for a bit.

The triangular numbers below 100 are 1, 3, 6, 10, 15, 21, 28, 36, 45, 55, 66, 78 and 91. Now, add any two consecutive triangular numbers, say, $3 + 6 = 9$, or $21 + 28 = 49$, or $66 + 78 = 144$.

What do you notice? It seems that each time we add two consecutive triangular numbers, we get a square number: $9 = 3^2$, $49 = 7^2$ and $144 = 12^2$. And if you check, you'll see it happens everywhere along this list. What is going on here?

It's most easily seen by looking at diagrams. Let's add 21 and 28 by looking at the triangular numbers: $21 + 28 = 49 = 7^2$.

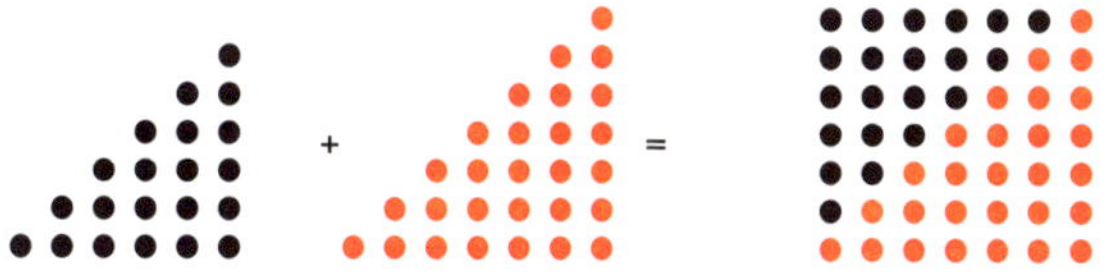

So a 7×7 square is clearly a triangle with base-6 and a triangle with base-7 added together. This will happen any time you add two consecutive triangular numbers.

If you want a really cool formula for this, write T_n for the nth triangular number and you get $T_{(n-1)} + T_n = n^2$.

EVERY BREATH YOU TAKE

While the air we breathe gives us the oxygen we need to live, it's actually 78% nitrogen. Oxygen makes up a mere 21% of the atmosphere. Argon makes almost all of the remaining 1% or so.

But it's not all about size – without oxygen we'd be cactus pretty quickly. When carbon dioxide concentration rose to a mere 0.04% or 400 parts per million (ppm) of our atmosphere, scientists expressed extreme concern that if it continues to a seemingly innocuous 0.05% (500 ppm) we will have wreaked havoc on our climate.

Back in chapter 74, I introduced many of you to the concept of the CD ...

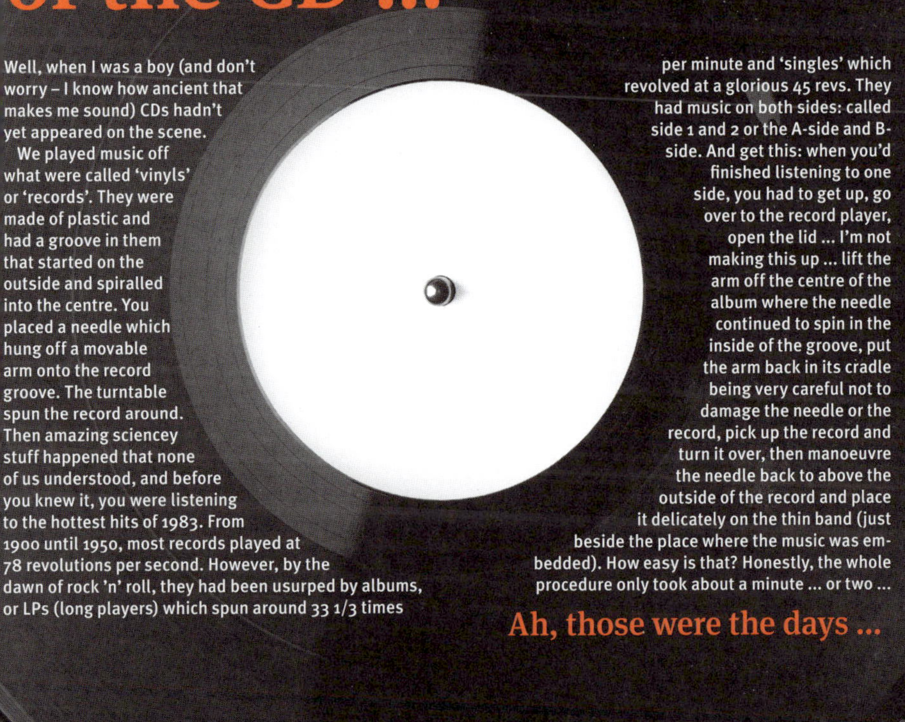

Well, when I was a boy (and don't worry – I know how ancient that makes me sound) CDs hadn't yet appeared on the scene.

We played music off what were called 'vinyls' or 'records'. They were made of plastic and had a groove in them that started on the outside and spiralled into the centre. You placed a needle which hung off a movable arm onto the record groove. The turntable spun the record around. Then amazing sciencey stuff happened that none of us understood, and before you knew it, you were listening to the hottest hits of 1983. From 1900 until 1950, most records played at 78 revolutions per second. However, by the dawn of rock 'n' roll, they had been usurped by albums, or LPs (long players) which spun around 33 1/3 times per minute and 'singles' which revolved at a glorious 45 revs. They had music on both sides: called side 1 and 2 or the A-side and B-side. And get this: when you'd finished listening to one side, you had to get up, go over to the record player, open the lid ... I'm not making this up ... lift the arm off the centre of the album where the needle continued to spin in the inside of the groove, put the arm back in its cradle being very careful not to damage the needle or the record, pick up the record and turn it over, then manoeuvre the needle back to above the outside of the record and place it delicately on the thin band (just beside the place where the music was embedded). How easy is that? Honestly, the whole procedure only took about a minute ... or two ...

Ah, those were the days ...

A tennis court is 78 feet long and 36 feet wide. But often that's not big enough for me to be able to land the ball in it!

LORDS-A-LEAPING!

Because 78 is the 12th triangular number, if you gave your true love all the gifts of the 12 days of Christmas you'd be handing over:

12 (drummers drumming) + 11 (pipers piping) + 10 (lords-a-leaping) + 9 (ladies dancing) + 8 (maids-a-milking) + 7 (swans-a-swimming) + 6 (geese-a-laying) + 5 (go-old riiiiiings) + 4 (calling birds) + 3 (French hens) + 2 (turtle doves) + 1 (partridge in a pear tree) = 78 gifts.

If anyone knows where I can grab a partridge in a pear tree, please drop me a line.

TAROT ET AL

Those of you into tarot (and I guess there must be some!) would know that there are 78 cards in a standard tarot deck – divided into 22 major and 56 minor 'arcana'.

Pinochle has 48 cards in the deck. Euchre 24. Five Hundred has 43.

The standard Spanish Deck or *Baraja Espanola* for Alcalde, Brisca, el Mus and so on requires 40 cards.

Uno has 108. And if you're keen on a quick game of the hugely popular 4-person *Austrian* card contest *Bauernschnapsen*, you'll be needing 20 cards. Any *Bauernschnapsen* freaks out there would not need me to tell them it derives from the German game *Sechsundsechzig* which means 66.

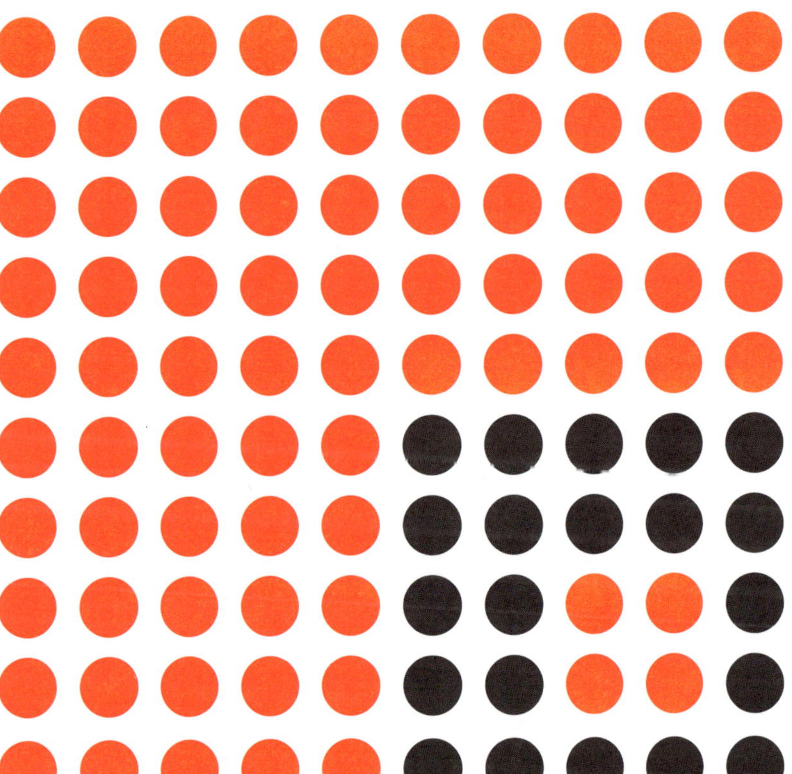

79

In the year 79 AD, the volcano Mount Vesuvius erupted, destroying the Roman cities Pompeii and Herculaneum killing an estimated 16,000 people.

Today some 3,000,000 people live near Vesuvius, making it the most densely populated volcanic region on Earth.

Here's one for you: $79 = 2^7 - 7^2$. Nifty, hey?

COUSIN PRIMES

We've met the twin primes, primes of the form $(p, p + 2)$ like (3, 5), (5, 7), (11, 13) and so on.

Well, when a pair of primes are separated by 4, not 2, they are called 'cousin primes'. So because 79 and 83 are both prime (79, 83) are cousin primes.

As of January 2006, the largest known pair of cousin primes are $(630{,}062 \times 2^{37{,}555} + 3,\ 630{,}062 \times 2^{37{,}555} + 7)$ which are each 11,311 digits long.

QUIZ QUESTION: Find all the cousin primes from 1 to 100. ANSWER AT THE BACK OF THE BOOK

On the topic of 79's primacy, it is also a happy prime and because when you write the prime number 79 backwards you get 97 which is also prime, we call 79 an 'emirp' (the word prime written in reverse).

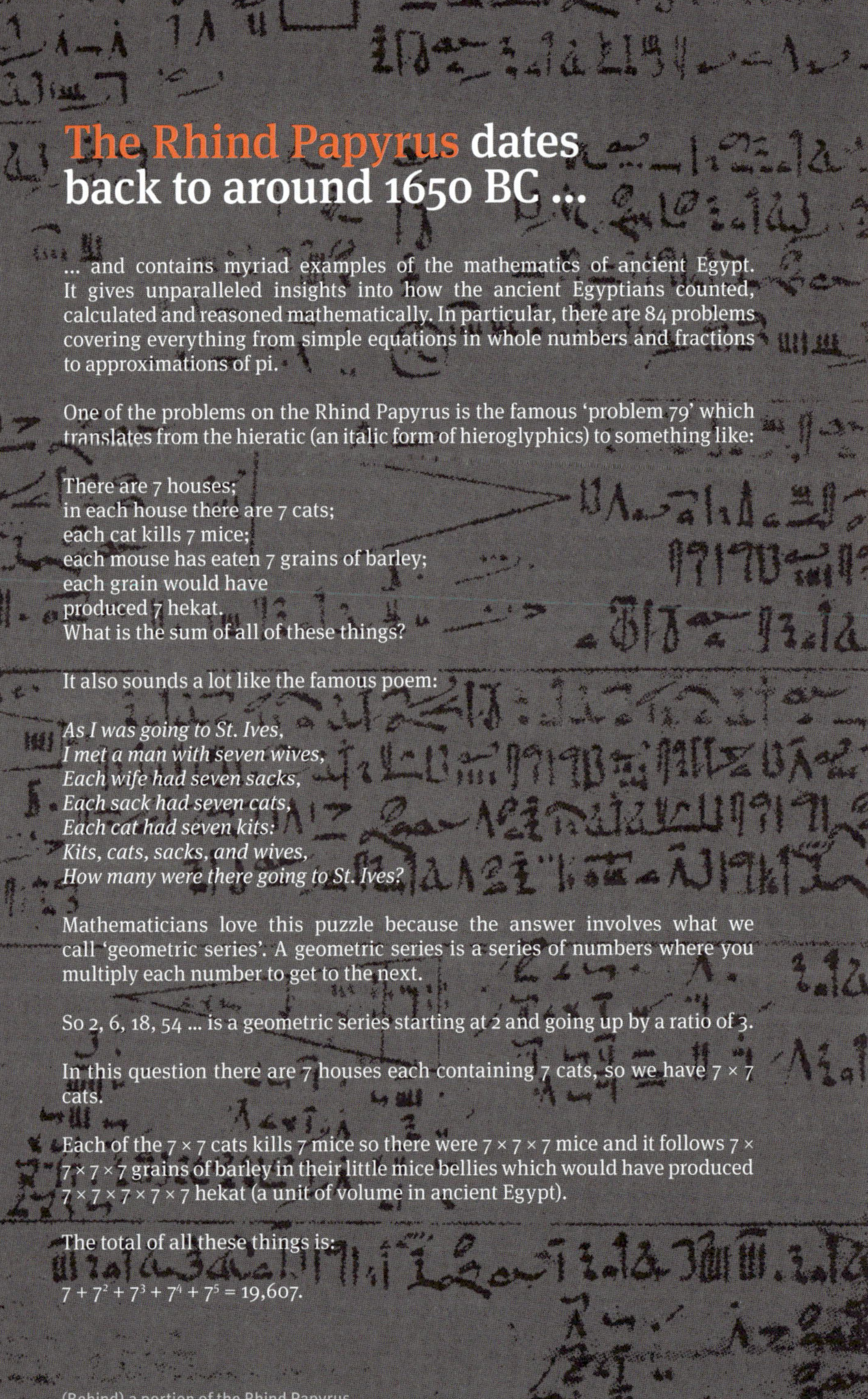

The Rhind Papyrus dates back to around 1650 BC ...

... and contains myriad examples of the mathematics of ancient Egypt. It gives unparalleled insights into how the ancient Egyptians counted, calculated and reasoned mathematically. In particular, there are 84 problems covering everything from simple equations in whole numbers and fractions to approximations of pi.

One of the problems on the Rhind Papyrus is the famous 'problem 79' which translates from the hieratic (an italic form of hieroglyphics) to something like:

There are 7 houses;
in each house there are 7 cats;
each cat kills 7 mice;
each mouse has eaten 7 grains of barley;
each grain would have
produced 7 hekat.
What is the sum of all of these things?

It also sounds a lot like the famous poem:

As I was going to St. Ives,
I met a man with seven wives,
Each wife had seven sacks,
Each sack had seven cats,
Each cat had seven kits:
Kits, cats, sacks, and wives,
How many were there going to St. Ives?

Mathematicians love this puzzle because the answer involves what we call 'geometric series'. A geometric series is a series of numbers where you multiply each number to get to the next.

So 2, 6, 18, 54 ... is a geometric series starting at 2 and going up by a ratio of 3.

In this question there are 7 houses each containing 7 cats, so we have 7×7 cats.

Each of the 7×7 cats kills 7 mice so there were $7 \times 7 \times 7$ mice and it follows $7 \times 7 \times 7 \times 7$ grains of barley in their little mice bellies which would have produced $7 \times 7 \times 7 \times 7 \times 7$ hekat (a unit of volume in ancient Egypt).

The total of all these things is:

$7 + 7^2 + 7^3 + 7^4 + 7^5 = 19{,}607$.

(Behind) a portion of the Rhind Papyrus.
Source: Public Domain

Charles Babbage, just chillin'.
Source: Public domain

The inventor of the Analytical Engine, Charles Babbage, died at the age of 79.

Although the Engine never progressed beyond detailed drawings, it was similar to a present-day computer. Babbage never built an operational, mechanical computer, but his design concepts have been proved correct and recently Babbage's computer was built following his own criteria. After his death, a British committee asked to report on the feasibility of the design commented that 'its successful realisation might be an epoch in the history of computation equally memorable with that of the introduction of logarithms'. Can I also give a shout out here to the amazing Ada Lovelace who worked with Babbage. She may well have written the first ever computer program. Go sister!

08

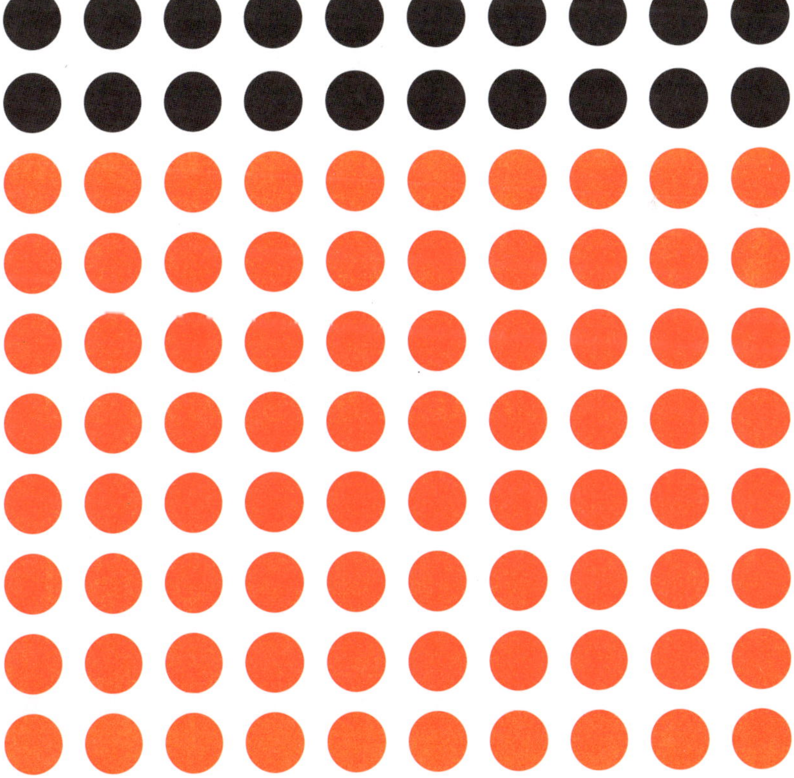

Eighty is a Harshad number. In telling you this I'm helping you answer the Quiz Question in chapter 84 where we explain what Harshad numbers are. Then again, it's a brute of a Quiz Question!

THE MAGIC STAR OF DAVID

A variation on the magic square is the magic star of David. There are 80 different magic stars of David that arrange the numbers 1 to 12 in grids like the ones below.

There are 6 rows and each entry is involved in 2 rows so it should be obvious that if all rows add up to the same number, that number will be [(1 + 2 + 3 + 4 + 5 + 6 + 7 + 8 + 9 + 10 + 11 + 12) + (1 + 2 + 3 + 4 + 5 + 6 + 7 + 8 + 9 + 10 + 11 + 12)] divided by 6, giving us a magic constant of 26 for our stars.

QUIZ QUESTION: Can you solve these magic stars of David? ANSWER AT THE BACK OF THE BOOK

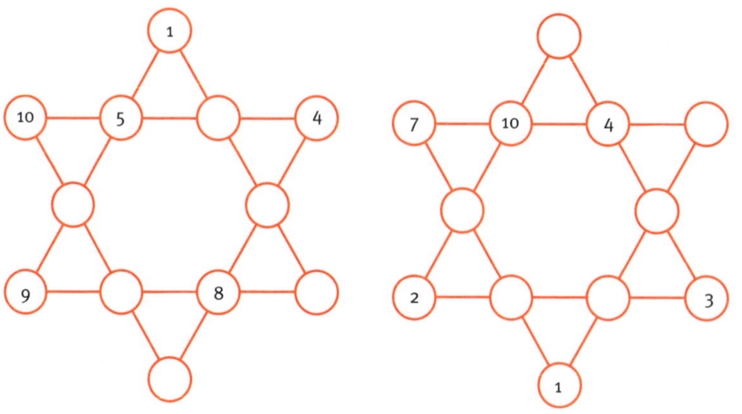

Parrots can live for up to 80 years – Polly want a cracker? I said, 'POLLY WANT A CRACKER?'

The Roman Colosseum was completed in the year 80 AD, which was good news for cats, but very bad news for Christians.

PSST! WANNA BUY 1.5M DIAMONDS?

On a clear night with no Moon your eyes can see a match being lit 80 kilometres (50 miles) away.

On the topic of lighting stuff (nothing to do with 80 but just too cool to leave out of this book) did you know that every second a burning candle produces one and a half million microscopic diamonds?

But before you go rushing out to buy a thousand candles and a net, the nano-particle gems are burnt away into carbon dioxide almost instantly.

BACK IN THE GAME

You often hear of sporting teams retiring a great player's number when they call it a day, but 6 games into the 2004 season, the Seattle Seahawks did something very rare. Legendary NFL wide receiver Steve Largent gave permission for his number 80 to be 'unretired' so the newly recruited fellow legend Jerry Rice could keep wearing the number he had worn in his time at Oakland and San Francisco.

William of Orange, circa 1582–
Source: Public Domain

The Dutch Revolt or Eighty Years War went for ... you guessed it ... 80 years.

It began in 1568 as an uprising by among others William of Orange against Philip the Second who despite being Spanish ran the Habsburg Netherlands.
It ended in 1648 when the Dutch Republic was finally recognised as an independent country.

You can only assume that no Dutch man or woman lived through the entire war.

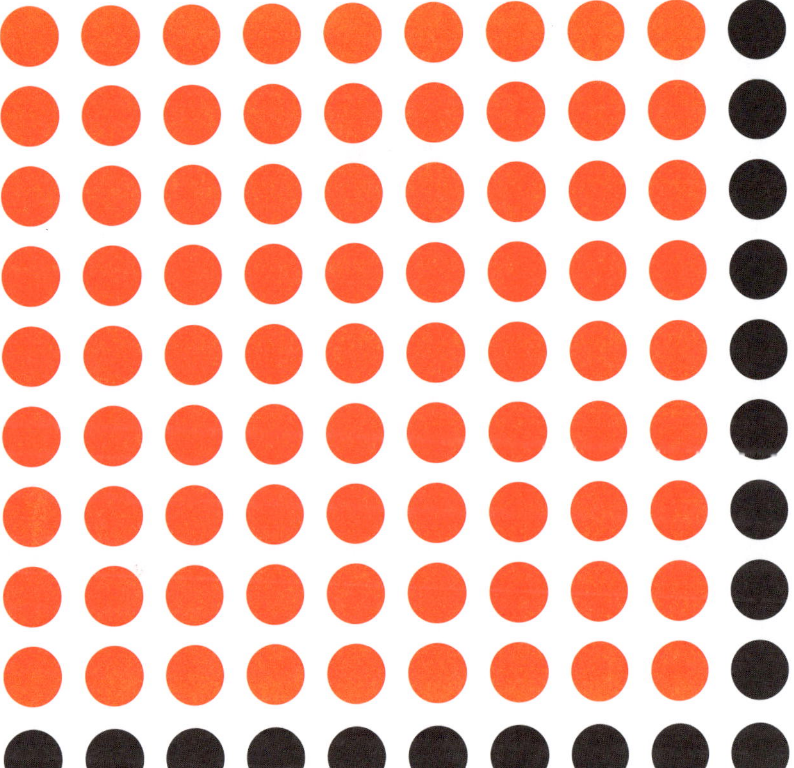

81

MR BURNS

The most easy to hate billionaire on *The Simpsons*, Charles Montgomery Plantagenet Schicklgruber Burns, tells Homer in the episode 'Simpson and Delilah' that he is 81 years old. Then again in the 'Who Shot Mr Burns?' episode he appears to be 104, another time he claims to be 123 and on one occasion when he enters his ATM PIN in answer to his age he types 4 digits which means he's possibly over 1000. Regardless, I'm sure when he hears he has made it into this book he will consider that 'excellent'.

9 x 9 MAGIC

Because 81 is a square it's possible to make 9×9 magic squares with the numbers 1 to 81. For magic squares see chapters 9, 13, 34, 64 and 65. There really is no excuse for not knowing what one is now!

But when it comes to asking, 'How many different 9×9 magic squares are there?' we have no idea. There are 880 different 4×4 squares, 275,305,224 types of 5×5 and for 6×6 we estimate there are a mind-boggling 17,745,000,000,000,000,000 or so squares. We simply can't calculate beyond that for magic squares size 7×7 or higher.

But let's just try one, hey?

Start by working out what all the columns in the diagram opposite add up to. Well, we know that the entire magic square contains the numbers 1 to 81, so all the squares in the magic square add up to:

$$1 + 2 + 3 + 4 + \ldots + 78 + 79 + 80 + 81$$

This looks daunting, but rearrange the order of the numbers and look at the sum like this:

$$1 + 81 + 2 + 80 + 3 + 79 + 4 + 78 + \ldots + 40 + 42 + 41$$

... you have to be a little bit careful at the end and make sure you don't miss out on the 41.

Now, $1 + 81 = 82$ but so does $2 + 80$ and $3 + 79$ and $4 + 78$ all the way up to $40 + 42$, we have 40 of these pairs that add up to 82 and the number 41 left over. So:

$$1 + 2 + 3 + \ldots + 81 = (1 + 81) + (2 + 80) + (3 + 79) + \ldots + (40 + 42) + 41$$
$$= 40 \times 82 + 41 = 3321$$

There are 9 rows and they are all equal so they must each add up to $\frac{3321}{9} = 369$.

Here is one of the 'who knows how many' 9×9 magic squares with constant 369.

1	16	51	30	45	77	56	71	22
41	47	61	67	73	9	15	21	35
69	75	8	14	20	34	40	46	63
13	19	36	42	48	62	68	74	7
53	32	64	79	58	12	27	6	38
81	60	11	26	5	37	52	31	66
25	4	39	54	33	65	80	59	10
29	44	76	55	70	24	3	18	50
57	72	23	2	17	49	28	43	78

EIGHTY-ONE

is the only number whose square root is equal to the sum of its digits (apart from 0 and 1). In the case of 81, 8 + 1 = 9 and 9 is the square root of 81. It's both square *and* heptagonal, which somehow seems like a contradiction in terms.

KEEP DRIVING

If Stephen King's 2011 story 'Mile 81' teaches us one lesson it's this – if you see a broken down wagon at a rest stop keep driving ... because if you don't, the wagon may well eat you!

HELLS ANGELS

Eighty-one is the symbolic number for the Hells Angels Motor Cycle Club. H is the 8th letter in the alphabet and A is the first hence HA – 81. I think that sounds eminently sensible – but to be honest even if I thought it was pretty silly I wouldn't be stupid enough to say so in print.

When we write out the fraction $\frac{1}{81}$ as a decimal we get 0.012345679 012345679 012 ... and so on. That is, the numbers 0 to 9 without 8, repeating endlessly. How could that happen?

Here's the proof that 0.012345679 0123456789 012345679 ... = $\frac{1}{81}$

Let x = 0.012345679 012345679 012345679 0 ...

Then $10x$ = 0.12345679 012345679012345679 0 ...

and $100x$ = 1.2345679 012345679 012345679 0 ...

and $1000x$ = 12.345679 012345679 012345679 0 ... and so on.

Each time we multiply by 10 the decimal place moves across one more spot to the right.

Keep going and eventually you'll get:

$1{,}000{,}000{,}000x$ = 12,345,679.012345679 012345679 0 ...

But we started with x = 0.012345679 012345679 012345679 0

and because the numbers after the decimal point are now all exactly the same for each of these terms when we subtract we just get:

$1{,}000{,}000{,}000x - x$ = 12,345,679.012345679 0 ... – 0.012345679 0 ...

So $999{,}999{,}999x = 12345679$

dividing each side by 999,999,999 we get:

$$x = \frac{12{,}345{,}679}{999{,}999{,}999} = \frac{1}{81}$$

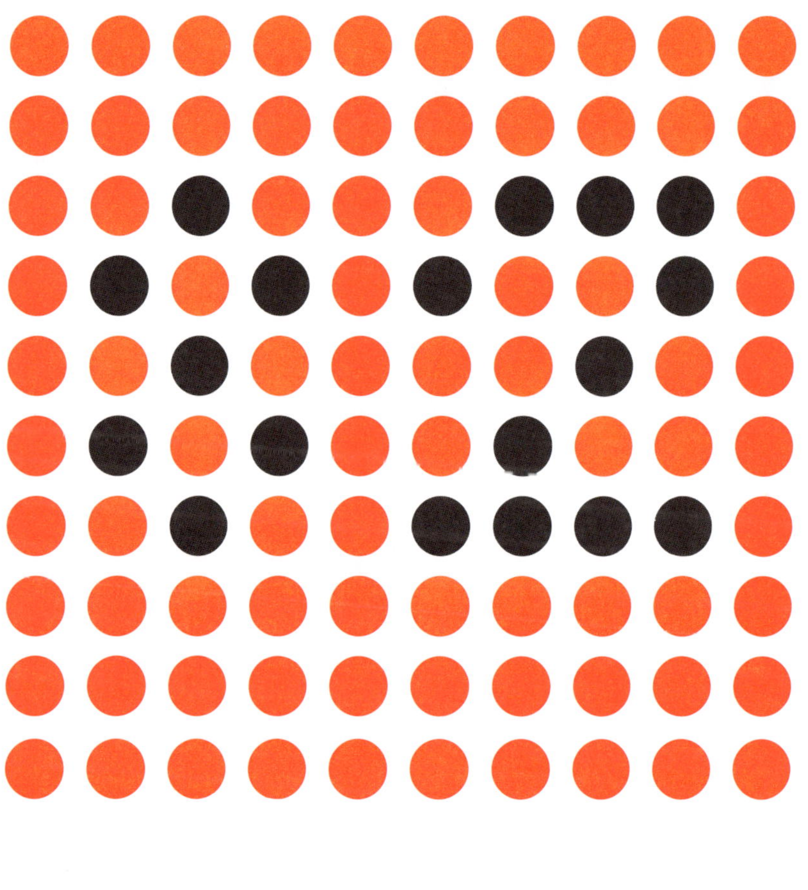

82

US senator Robert F. Kennedy (brother of assassinated American President John F. Kennedy) launched his presidential campaign on 16 March 1968. The campaign lasted only 82 days as the Democratic hopeful was assassinated on 6 June of that year.

King Sobhuza of Swaziland ruled for 82 years having taken over at the age of only four months. According to Swazi authorities, he married 70 wives who, over a 50-year period, bore 210 children. At his death, he had over 1000 grandchildren!

VERY SUPERSTITIOUS ...

It was revealed in mid-2014 that Sydney's soon-to-be-built tallest residential tower will go up to level 82 but only have 66 floors. How could this be? Many Asian cultures are superstitious about the number 4 – the word for '4' sounding similar to the word for 'death' in some languages. So the 66-storey building has no levels 4, 14, 24, 34, 40, 41, 42, ... 49, 54, 64 or 74.

It all seems a little strange to me – but no stranger than the lack of a 13th floor in most hotels.

PLUMBUM

Eighty-two is the atomic number of the soft malleable metal lead (Pb). Lead has a melting point of 327.46° Celsius (621.43° F) and if you're not happy having melted lead and actually want to boil it, you'll have to crank your system up to a tasty 1749° C (3180° F) – suffice to say you won't want to be wearing a jumper under your protective clothing.

When a group of lead atoms hang out together in a crystal structure, they do it as a face-centred cube.

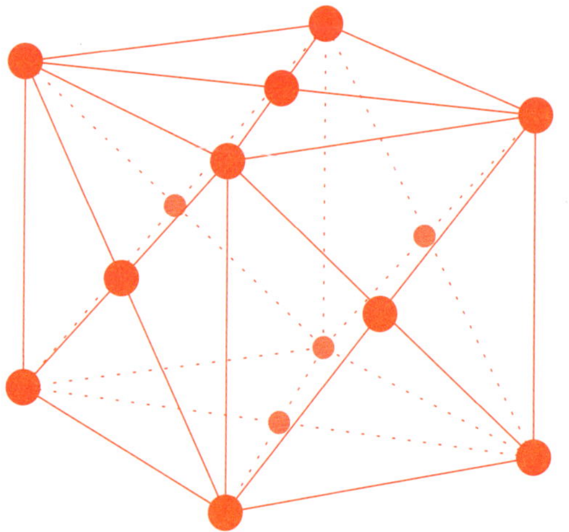

Why, you may well ask, is lead given the symbol Pb? Well, lead has been used way back in our history – metallic lead beads have been found in what is now Turkey that are over 8000 years old.

The Romans loved nothing more than smelting a bit of lead – they used vast amounts for building, piping water and all manner of stuff. So the symbol Pb comes from the Latin word *plumbum* which means 'lead' and gives us the word 'plumbing'.

I was particularly happy when I found this out while researching this book – one because I love learning new things and two, because it's not every day you get to write 'plumbum'.

A standard dartboard has 82 sections
– the numbers 1 to 20 are broken into 4 sections and the inner ring and bullseye.

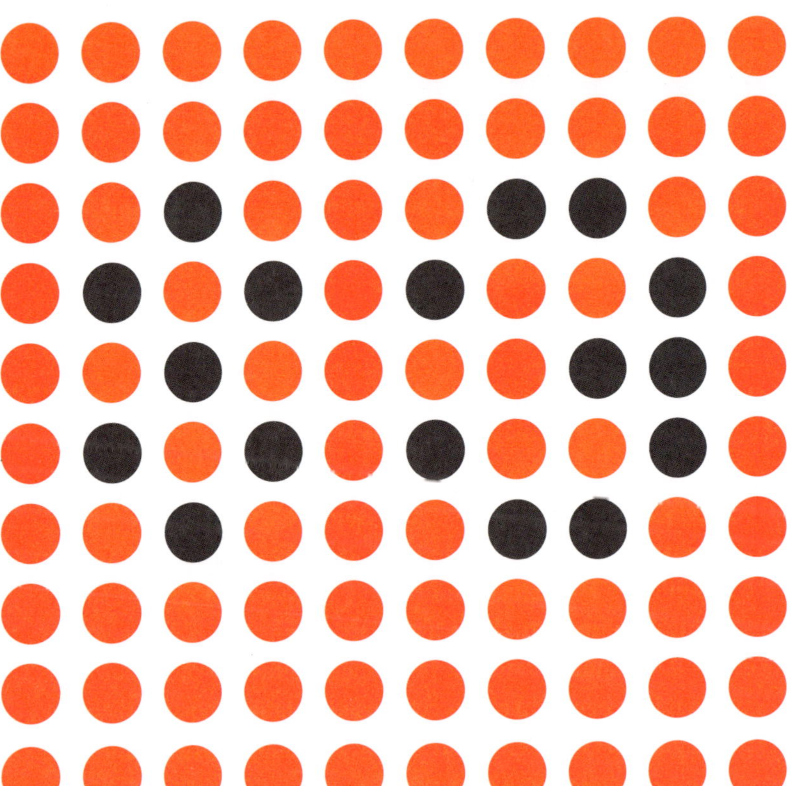

83

EUCLID'S INFINITE AWESOMENESS

Eighty-three is prime. About 2300 years ago, the famous Greek mathematician Euclid proved that there are an infinite number of primes in Proposition 20 of Book IX of his giant work *Elements*. The big guy did it like this:

Let's say there is only a finite number of primes 2, 3 and 5 and then they stop.

Now think about the number: $(2 \times 3 \times 5) + 1 = 31$.

Let's divide 31 by each of the primes on the list of all primes.

Clearly it isn't divisible by 2, because 2 divides into $2 \times 3 \times 5$ and you get a remainder of 1 that comes from the + 1 at the end.

Similarly, it isn't divisible by 3 or 5, because each time the division will still leave a remainder of 1.

So if we assume that the list of all primes is 2, 3 and 5 we can create a new number 31 that isn't divisible by any of 2, 3 or 5. So, 31 is either a new prime or a composite number divisible by a new prime that isn't on our list. Either way, that supposedly complete list of primes wasn't actually complete.

Anytime someone says 'there are only a finite number of primes – here is the list' you could just do what we've done here. Multiply them all, add 1 an,d by considering the new number, you've shown that list of primes was incomplete.

So there can be no finite list of primes, ever. There must be an infinite number of primes.

Great work, Euclid – you truly rock.

At the same time, we know that there must always be a largest known prime number. To catch the largest we've currently found, see chapter 31.

We're about to find out why 83 is a 'safe' prime. But at the same time as being safe because 83 and $83 + 6 = 89$ are both prime, we say that 83 is a sexy prime! If you're struggling to get excited by this 'sexy prime', don't worry – the 'sex' in 'sexy prime' refers to the Latin word for six.

DANGER, DANGER!

The innocent Will Robinson, the clunky robot and the hapless Dr Smith got into 'Danger, Danger' on 83 episodes of *Lost in Space*. Every one of which was better than the incredibly weak 1998 movie *Lost in Space* starring Matt LeBlanc.

VANUATU

is an archipelago of 83 islands lying between New Caledonia and Fiji in the Southwest Pacific. The main island is Efate, where Port Vila is located.

IN 1812 ...

8-year-old Zerah Colburn from Vermont, US, visited Europe to demonstrate his skills. He could instantly give the product of 2 numbers each of 4 digits. When asked for the factors of 171,395, he quickly gave 5, 7, 59 and, yep, 83. Handy trick.

SAFE PRIMES

The number 83 is the sum of 3 consecutive primes and the sum of 5 consecutive primes. Can you work out each of the sums?

$2 \times 83 + 1 = 167$ which is prime, so 83 is a Sophie Germain prime (see chapter 89).

But 83 is also a 'safe prime'. A prime is safe if it can be written in the form $2p + 1$ where p is a prime.

So $83 = 2 \times 41 + 1$ and 41 is prime, so 83 is prime.

Safe primes might sound a bit naff but they actually have important applications in cryptography.

QUIZ QUESTION: There are 7 safe primes under 100. We know that 83 is one of them – find the other 6. Apart from the first two that are under 10, what do the other four have in common with 83? ANSWER AT THE BACK OF THE BOOK

There are 83 'dry' towns and villages in Alaska

... and, in addition, Fairbanks is a dry town for moose, where it's illegal to feed a moose any alcoholic beverage. Apparently they can't hold their liquor.

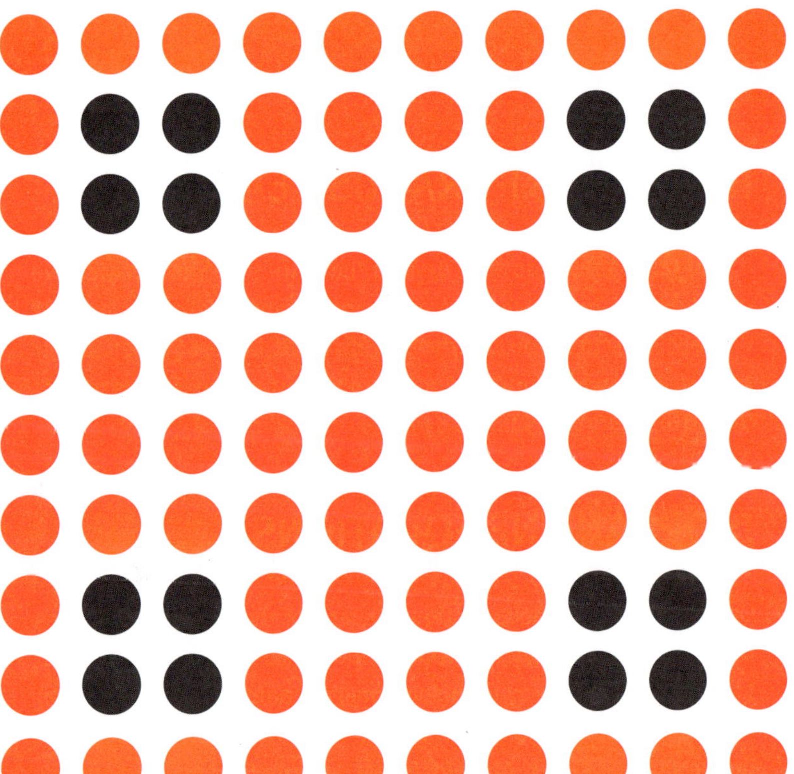

84

HAPPY FIRST URANUS YEAR!

A year on Uranus (one orbit around the Sun) is 84 Earth years. The very talented botanist, musical composer and telescope designer William Herschel who discovered the third largest planet in our solar system, passed away tantalisingly short of one Uranus year at 83 (Earth) years and 9 months of age.

DIOPHANTINE EQUATIONS

One of the most famous mathematicians of the ancient world was Diophantus, after whom Diophantine equations (equations that have whole numbers as their solutions) are named (see also chapter 61). Diophantus was so maths-mad there is even a riddle or equation in the famous *Greek Anthology* that apparently appeared on his tombstone:

God granted him to be a boy for 1/6 of this life, and
adding 1/12 part to this, he clothed his cheeks with down.
He lit him the light of wedlock after 1/7 part,
and 5 years after his marriage he gave him a son.
Alas, late-born and wretched child! After attaining
the measure of half his father's life, chill Fate took him.
After consoling his grief by the study of numbers
for 4 years, Diophantus ended his life.

QUIZ QUESTION: Can you prove that the answer to this riddle is that Diophantus lived to be 84? ANSWER AT THE BACK OF THE BOOK

Opposite page: The title page of the 1621 edition of Diophantus' *Arithmetica*, translated into Latin by Claude Gaspard Bachet de Méziriac. Source: Wikipedia

DIOPHANTI ALEXANDRINI ARITHMETICORVM LIBRI SEX,

ET DE NVMERIS MVLTANGVLIS LIBER VNVS.

Nunc primùm Græcè & Latinè editi, atque absolutissimis Commentariis illustrati.

AVCTORE CLAVDIO GASPARE BACHETO MEZIRIACO SEBVSIANO, V.C.

LVTETIAE PARISIORVM,
Sumptibus SEBASTIANI CRAMOISY, via Iacobæa, sub Ciconiis.

M. DC. XXI.

CVM PRIVILEGIO REGIS.

THAT'S HARSH(AD)

Eighty-four is an abundant number (see chapter 12) but it is also a Harshad number.

A number that is divisible by the sum of its own digits is called a Harshad number.

So 84 is a Harshad number, because $8 + 4 = 12$ and $84 = 12 \times 7$.

Similarly, 1729 is a Harshad number because $1 + 7 + 2 + 9 = 19$ and $1729 = 19 \times 91$.

It's obvious that 1-digit numbers are Harshad, as are multiples of 10 – for example, for 50, $5 + 0 = 5$ and $50 = 5 \times 10$.

QUIZ QUESTION: Find all the Harshad numbers from 10 to 100. Warning – it will take quite a while ... but just imagine the sense of reward! ANSWER AT THE BACK OF THE BOOK

1984 AND ALL THAT

The year *Ninteen Eighty-Four* was immortalised by George Orwell in his book of the same name. Inspired by George's tome, David Bowie wrote a pretty cracking rock and roll hit called ... you guessed it, '1984', which featured on the album *Diamond Dogs*. And British movie director Michael Radford, inspired by George Orwell a little bit more than David Bowie I'm assuming, made a movie called ... you'll never guess ... *1984* which was released in ... go on ... just have a wild stab in the dark ... oh, how did you know ... 1984.

In 1984, I failed to qualify for the final of the men's 100m sprint. It was won by Carl Lewis in a time of 9.99 seconds.

Don't feel too sorry for me, I'm not that quick over 100 and I didn't enter.

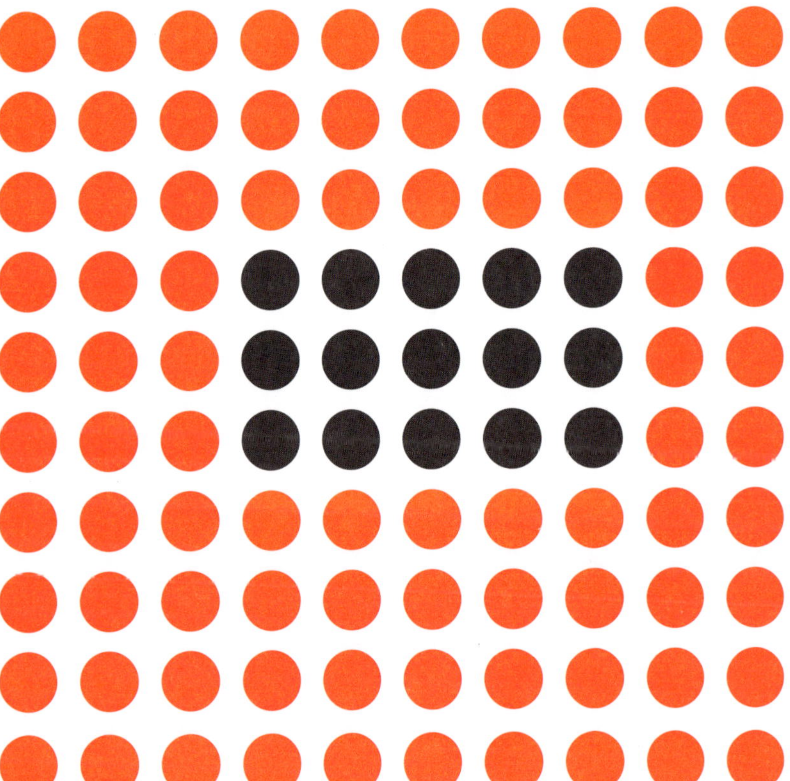

85

NECKTIES

One of my favourite ever examples of applying hardcore mathematics to an everyday problem involves the number 85.

Consider tying a necktie. For all the fancy stuff you might try to do, often committed to memory with catchy little mnemonics like 'the fox comes out of his hole, sees the rabbit, chases it round the tree' and so on, any way that you might tie a necktie comes down to these simple observations:

At any given time in tying a necktie, the end of the tie can be moved to the left, back through the centre, or to the right.

If the end of the tie is already in a position, it can't go to that position again (say if it's on the right, it can't go to the right again, it must go to the centre or the left).

And ... every combination must finish Left, Right, Centre or Right, Left, Centre.

So every possible way you might want to tie a necktie can essentially be summarised as 'LRCRCLRLRC end', or similar.

In their groundbreaking mathematico-fashionista masterpiece *The 85 Ways to Tie a Tie – The Science and Aesthetics of Tie Knots*, Thomas Fink and Yong Mao worked out the 85 different ways you could possibly entangle a standard-length necktie.

While most of the results were completely ugly and impractical, they did improve upon the 4 ways that were already known – the four in hand, the Windsor, the Half-Windsor and the Pratt – with a whopping 6 new practical knot styles and 2 massively complicated arrangements.

Who says mathematics doesn't touch the real world, hey stylemasters?

EVITA VS CLEOPATRA

The highest recorded number of costume changes ever performed in a Hollywood movie was 85, by Madonna in *Evita*. She also sported 39 hats, 45 pairs of shoes and 56 pairs of earrings in her role as Eva Peron. In setting the record she blitzed the previous mark of 65 costume changes held by the not-exactly-wardrobe-shy Elizabeth Taylor in *Cleopatra*.

We saw earlier that 65 can be written as the sum of two squares in two different ways.

$$65 = 8^2 + 1^2 = 4^2 + 7^2$$

QUIZ QUESTION: Try to find the two ways that 85 can be written as the sum of two squares. ANSWER AT THE BACK OF THE BOOK

FUNDAMENTAL PARTICLES

When a physicist says something is a 'fundamental particle' they mean it can't be broken into anything smaller that we know of – it is essentially a 'building block' of the universe.

Fundamental particles come in various guises – some, like electrons, you've probably heard of, others with pretty cool names might be new to you – quarks, neutrinos, photons, muons, gluons and of course our amazing discovery from 2012, the Higgs boson. These particles and their anti-particles make up all of the 'stuff' in the universe.

Any guess as to how many fundamental particles make up our observable universe is exactly that, a guess, but the best guesses going around seem to range between 10^{80} (1 followed by 80 zeros) up to 10^{85} (1 followed by 85 zeros).

SUPER SEVENTHS

$102^7 = 12^7 + 35^7 + 53^7 + 58^7 + 64^7 + 83^7 + 85^7 + 90^7$ ... trust me.

So 85 is part of this equation which shows that 102 is the smallest number whose 7th power is the sum of 8 other 7th powers.

98

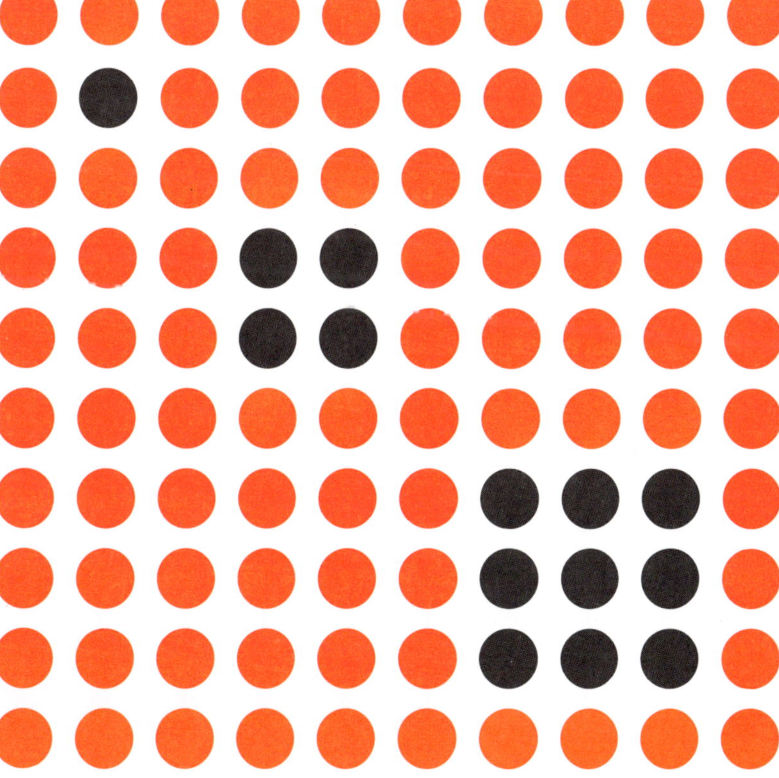

SEMIPRIME

The number 86 is not prime, but when we break it down into all of its factors we get $86 = 2 \times 43$ and because it only has two prime factors we call 86 'semiprime', almost prime or biprime. I'll stick with semiprime here.

Notice that $85 = 5 \times 17$, $86 = 2 \times 43$ and $87 = 3 \times 29$ where 2, 3, 5, 17, 29 and 43 are all prime.

QUIZ QUESTION: 85, 86 and 87 are the second run of three consecutive semiprimes. Find the first. ANSWER AT THE BACK OF THE BOOK

PADOVAN, YOU DA MAN

Check out this beautiful diagram:

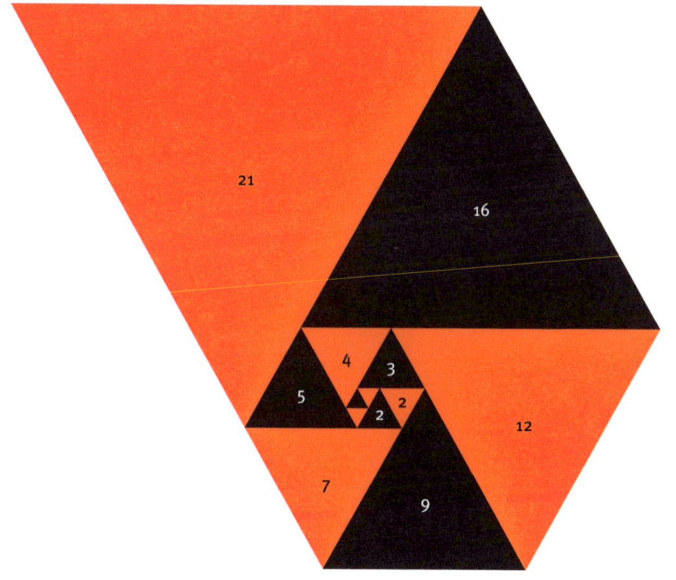

The triangles spiral out from the centre and always fit together but they get bigger and bigger.

The lengths of the sides of the triangles are 1, 1, 1, 2, 2, 3, 4, 5, 7, 9, 12, 16, 21 …

And when we look at this diagram that expands out a bit we see they continue:

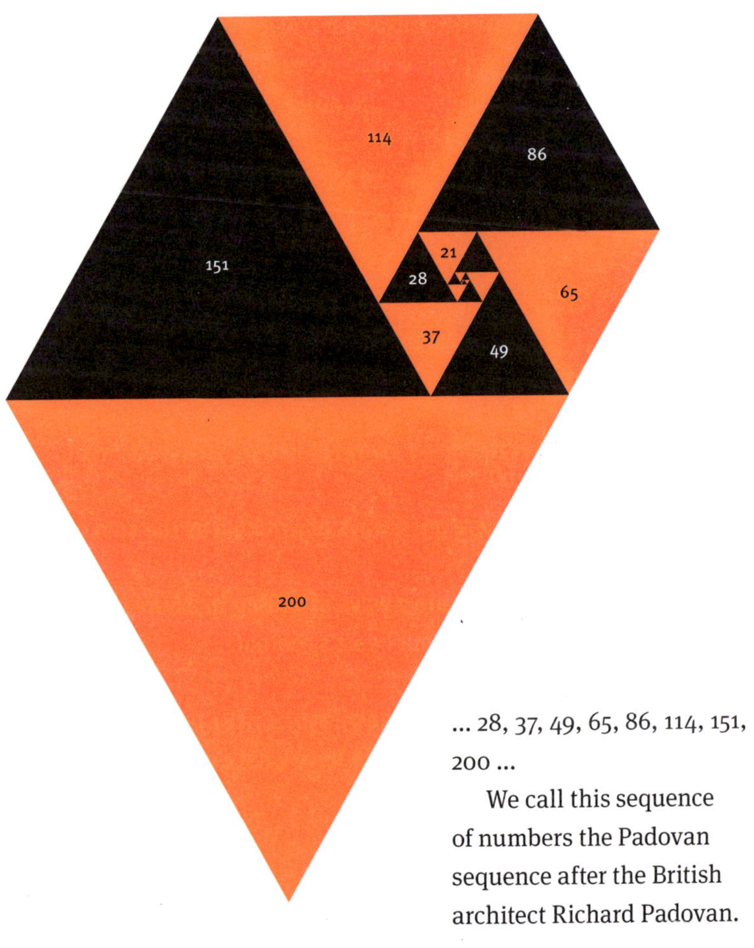

… 28, 37, 49, 65, 86, 114, 151, 200 …

We call this sequence of numbers the Padovan sequence after the British architect Richard Padovan.

PADOVAN'S RECURRENCE RELATION

We learnt about 'recurrence relations' when we looked at the Pell numbers back in chapter 70.

Convince yourself that the Padovan numbers obey the recurrence relation:

$$P(0) = P(1) = P(2) = 1 \text{ and } P(n) = P(n-2) + P(n-3)$$

This is another one of those examples of how different things in maths can be related that you would never have guessed.

The Padovan number $P(n)$ is the number of ways of writing the number $(n + 2)$ as the ordered sum of 2s and 3s.

For example, when $n = 6$, we get $P(6) = 4$, and there are 4 ways to write 8 $(6 + 2)$ as the sum of 2s and 3s where the order matters:

$$2 + 2 + 2 + 2;\ 2 + 3 + 3;\ 3 + 2 + 3;\ 3 + 3 + 2$$

QUIZ QUESTION:
P(5) = 3 so show that there are 3 ways of writing 7 as the sum of 2s and 3s
P(7) = 5 so show that there are 5 ways of writing 9 as the sum of 2s and 3s
P(8) = 7 so show that there are 7 ways of writing 10 as the sum of 2s and 3s ... and so on unto your heart's content.

QUIZ QUESTION: 2^{86} = 77,371,252,455,336,267,181,195,264 is the largest power of 2 that we know of to have what quality?

ANSWERS AT THE BACK OF THE BOOK

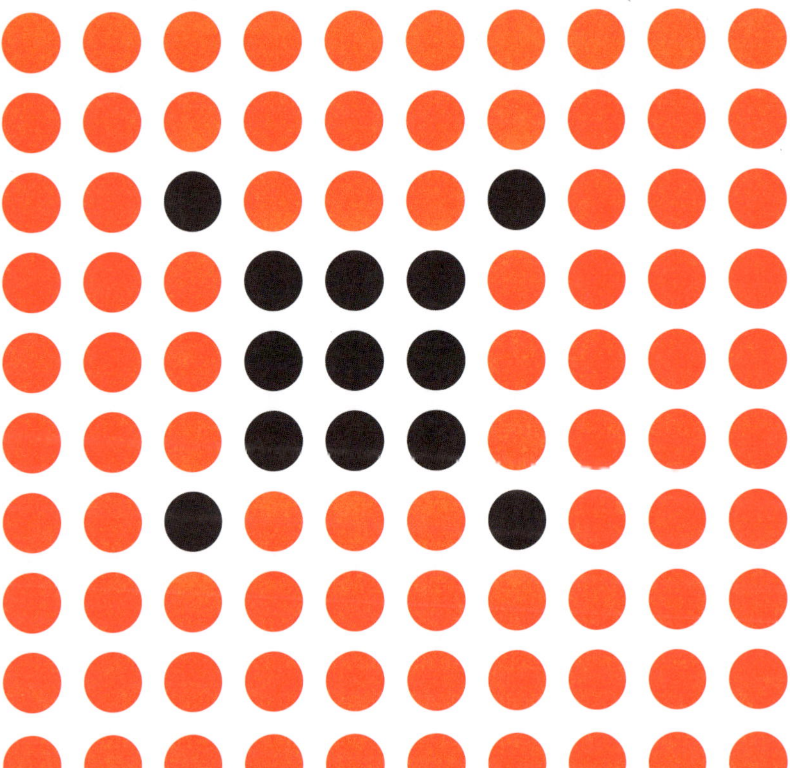

87

ARISTARCHUS OF SAMOS

Aristarchus was a Greek astronomer who genuinely had his gear together way back around 270 BC. Using a modified sundial called a *skaphe*, he could measure both the direction and height of the Sun.

As a result of some pretty neat maths, he calculated that the angle between the Sun, the Earth and the Moon was 87 degrees during a half-Moon.

Look, it's unlikely that he said 'heaps' or 'miles' for obvious reasons. In fact, the angle is greater than 89 degrees and the Sun is about 400 times further away from Earth than the Moon is – Aristarchus only guessed that the Sun was about 18 to 20 times as far away. Nonetheless, a beautiful piece of reasoning for the time.

Aristarchus also thought that the Earth went around the Sun and not vice versa, which is pretty much accepted now, but was way controversial back then.

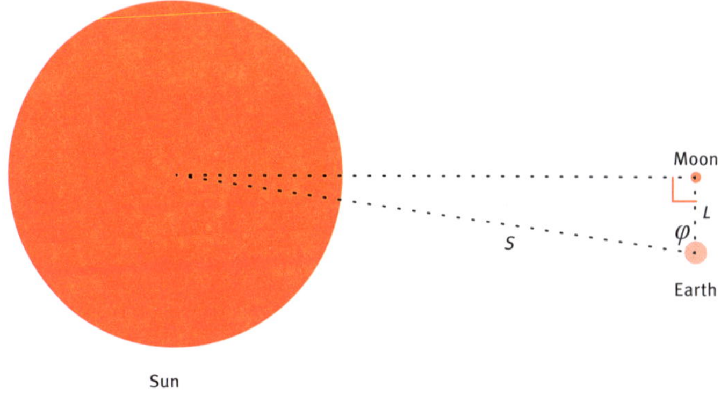

GETTYSBURG

There were 87 years between two events of major importance in American history – the signing of the American Declaration of Independence in 1776 and the Battle of Gettysburg in 1863.

But it took the oratorical skills of Abraham Lincoln to know that 'Four score and seven years ago' would sound so much better than '87 years back' or 'ages ago' when he gave the now famous address at the dedication of the Soldiers' National Cemetery in Gettysburg.

NOT SO LUCKY?

Eighty-seven is the 21st lucky number. Ironically, 13 is also a lucky number (see chapter 25 for a refresher on lucky numbers).

Someone who is likely to dispute that 87 is indeed lucky would be an Australian cricketer. Cricket in Australia holds 87 as a spectacularly unlucky score and is referred to as 'the devil's number'. Poor old 87's cause is not helped by the fact that it's 13 runs short of a century and whenever an Australian cricketer gets to 87 runs in an international game you can almost hear the clenching of stomach muscles, teeth and other body parts around the ground.

But it turns out there is nothing for batsmen to be scared about – like practically every superstition, this one is a myth, according to renowned statistician Ric Finlay.

One day Ric had a bit of time on his hands (which is often the case with cricket statisticians) so he decided to pour over the entire history of Australian Test cricket innings (which again is often what cricket statisticians do when they've got some time – hey, don't knock it until you've tried it).

Ric found that in the history of the game, only 12 Australian players have been dismissed on Oz cricket's devil's number. More unlucky are the numbers 85 (22 batsmen dismissed), 86 (13 batsmen dismissed), 88 (20 batsmen dismissed) and 89 (19 batsmen dismissed).

Probably the greatest athlete in any sport, in any country, in any time, Sir Donald Bradman, averaged 99.94 runs when batting in Test cricket.

In fact, it turns out that the hype surrounding 87 has nothing to do with it being 13 runs short of 100 – that's just a coincidence. It was actually started by cricketing great Keith Miller.

In December of 1927, when Miller was only 10, he was at the Melbourne Cricket Ground watching 'The Don' play.

Keith's hopes of seeing his idol score 100 were dashed when he was bowled by Harry 'Bull' Alexander – for 87. Miller was scarred and always noted when a batsman got out for that score. When Miller himself went on to play for Australia, he told everyone his theory that 87 was unlucky and a sporting superstition was born.

But in a beautiful irony, in later life Keith Miller once looked back on that 1929 scorebook to remind himself of that fateful day.

Don Bradman had actually scored 89 ... the scoreboard was wrong!

The Don plays a shot.
Source: Public Domain

88

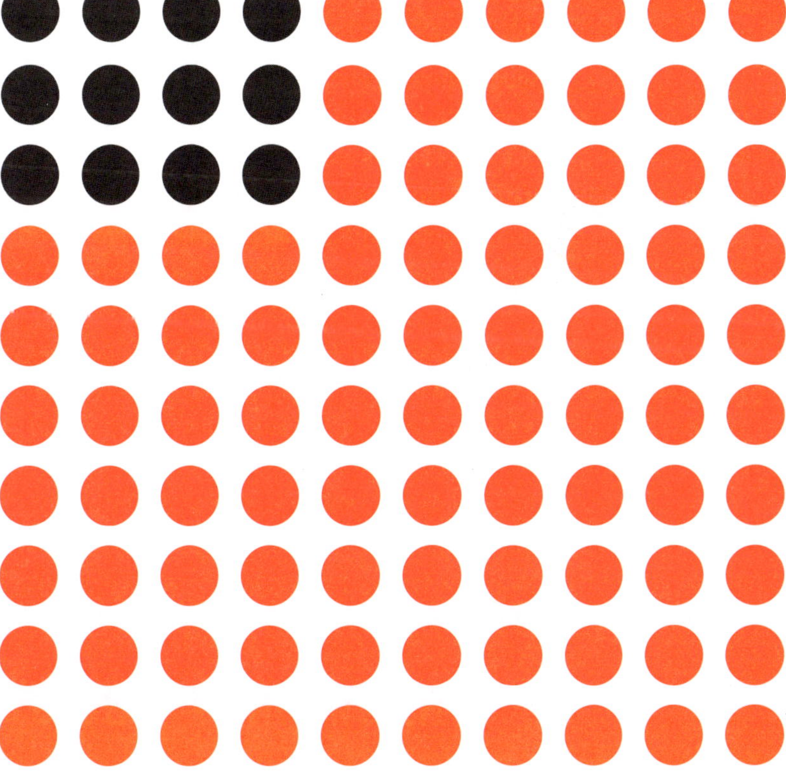

STARSKY AND HUTCH

Readers with no pop culture knowledge before the year 2004 may think that the only version of *Starsky and Hutch* is the film of that year featuring Ben Stiller as Starsky, Owen Wilson as Hutch and of course Snoop Dogg as the informant Huggy Bear.

Time to dust off the old DVD player or hop onto YouTube and brush up on some ancient history ... the TV version of Starsky (Paul Michael Glaser) and Hutch (David Soul), blowdried their hair, reached for their sunglasses and used a crowbar to get into their jeans 88 times in the 1970s.

And for an extra point, the street-wise supergrass Huggy Bear was originally played by Antonio Fargas!

I should point out two things – first, there was also a 70-minute TV pilot and some online entries suggest 92 not 88 episodes were made. Regardless, if you sit through 88 episodes you'll get a pretty good idea what *Starsky and Hutch* was all about.

SYMMETRIC MAGIC SQUARE

Rob Eastaway's great book *How Many Socks Make a Pair?* contains this little beauty:

25	18	51	82
81	52	15	28
12	21	88	55
58	85	22	11

... which is also 'magic' upside down, and in a mirror.

Radium has an atomic number of 88 and was discovered by the great Marie Curie and her husband Pierre in 1898

... the same year they also discovered Polonium (atomic number 84). Initially they discovered it in the form Radium Chloride but Marie Curie then isolated it in its metallic state in 1910.

Marie Curie was the first woman to win the Nobel Prize but was so awesome she won two ... in different sciences. In 1903, she won the prize for Physics with Pierre and Henri Becquerel and in 1911 she backed it up with the Prize for Chemistry.

She died in 1934 of cancer from exposure to radiation over her career including the X-ray machines she invented for use in World War I.

Marie Curie's Nobel Prize portrait. Source: Public Domain

In Chinese language chats, text messages, SMS and IM, the number 88 is used as a goodbye sign-off because in Mandarin 88 is pronounced 'ba ba' which phonetically resembles the English sounding 'bye bye'. Somewhat less innocently, 88 was also the Nazi code for 'Heil Hitler' with H being the 8th letter of the alphabet, therefore HH = 88.

A year on Mercury is only 88 days as it veritably races around the Sun.

BIRTHDAY PROBLEM REDUX

Back in chapter 23 we learnt about the famous and counter-intuitive result called the birthday problem. This paradox says that in a group of 23 people there's a better than even chance that there will be a shared birthday. Well, in a group of 88 people there is a better than even chance that *three* of them will share a birthday.

88 KEYS

A standard concert piano has 88 keys (52 white, 36 black) but it hasn't always been that way. The piano evolved from the harpsichord around 1700, and had 60 keys mimicking the harpsichord's 60 notes – 5 octaves at 12 notes each.

Over time this evolved to 84 keys (7 octaves) then in the 1880s one of the world's most famous manufacturers, Steinway & Sons, added 4 keys to give us the 88 we know today.

Not to be outdone, the Australian piano makers Stuart & Sons had manufactured the big momma of all pianos with 102 keys!

68

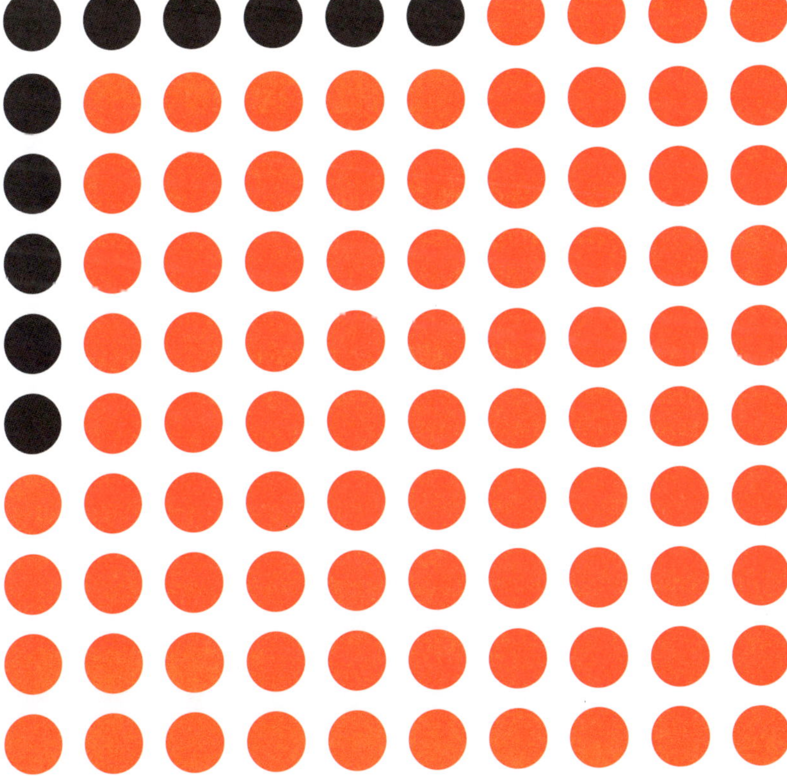

Each year around the anniversary of the Tiananmen Square massacre the Chinese Government censors the number 89 (along with the numbers 6 and 4) on search engines and Chinese social networking sites. Internet users are confronted with the message, 'According to the relevant laws and policies, the results of your search "89" cannot be displayed'. The move is intended to prevent people from accessing or spreading their memories of what happened on 4 June 1989 when an unknown number of people were killed during a military crackdown on pro-democracy protests in the centre of Beijing. The numbers are censored because of the date they represent 6 (the month of June), 4 (the date) and 89 (the year).

89 98 AND ALL THAT

Rob Eastaway has written some cracking books on mathematics and in *How Many Socks Make a Pair?* he shows that when you add a number to its reverse and keep doing that, you usually get quickly to a palindromic result. He gives the example 382:

382 + 283 = 665

665 + 566 = 1231

1231 + 1321 = 2552 a palindrome.

Similarly, 59 within 3 steps will become 1111.

But if you start with 89 + 98 = 187, you'll need 24 steps before finally getting to 1,801,200,002,107 + 7,012,000,021,081= 8,813,200,023,188 at which point I think you'd be feeling a mixture of relief and a tremendous sense of 'why did I bother doing that'?

It could have been worse. You could have started with 196, which we've tested for millions of steps and not yet found it reducing to a palindrome. As a result this procedure is called the 196-algorithm.

GERMAIN IN THE MEMBRANE

Sophie Germain was an amazing mathematician back in a time when it was particularly difficult for women to show off their mathematical skills. Women weren't allowed to attend Paris' École Polytechnique around 1800, so she borrowed lecture notes from men who were enrolled. She wrote mathematics that caught the attention of the great Lagrange, but did so under the man's name 'Monsieur LeBlanc'. When Lagrange wanted to meet this LeBlanc character he was shocked to discover that 'he' was a she. They became good friends and he mentored the brilliant Frenchwoman.

A Sophie Germain prime is defined as any prime number p for which $2p + 1$ is also prime.

So 5 is prime, and $2 \times 5 + 1 = 11$ is also prime. Therefore 5 is a Sophie Germain prime.

Similarly, 89 is a Sophie Germain prime because 89 is prime, as is $2 \times 89 + 1 = 179$. The primes so generated are called 'safe' primes (see chapter 83).

While this might look cute, but of no great help, Sophie used these sorts of primes to make early inroads into Fermat's Last Theorem (see chapter 67). Nice one, Soph.

In April 2012 the Sophie Germain prime:

$18{,}543{,}637{,}900{,}515 \times 2^{666{,}667} - 1$ was found ...

... it's over 200,000 digits long.

QUIZ QUESTION: **Eighty-nine is the last Sophie Germain prime between 1 and 100. Find the previous 9.** ANSWER AT THE BACK OF THE BOOK

BOY, GIRL, BOY, GIRL

Here's another cool thing about the Fibonacci numbers.

Imagine we have a row of chairs and boys and girls who want to sit in them. How many arrangements can we have as long as two boys don't sit next to each other (you know what boys are like!)

One chair, two arrangements.

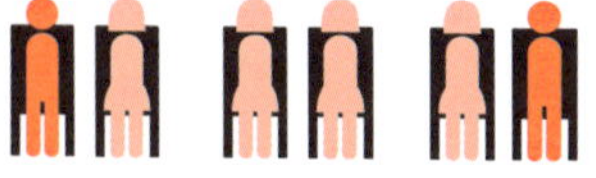

Two chairs, three arrangements.

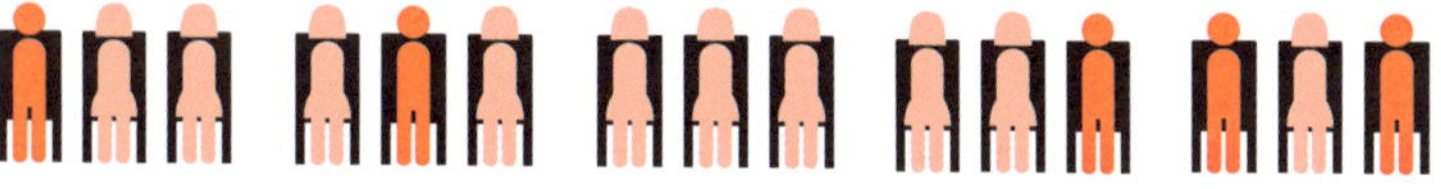

Three chairs, five arrangements.

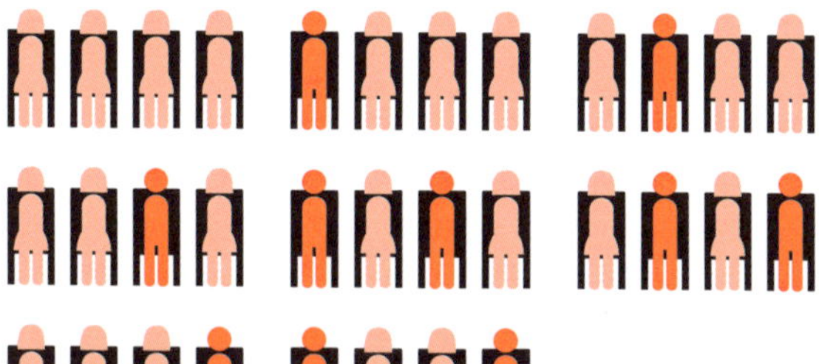

Four chairs, eight arrangements.

What do you notice about the answers 2, 3, 5, 8?

That's right, they're the Fibonacci numbers!

Why not convince yourself that when you have 5 chairs there are 5 + 8 = 13 ways and just take it from me that with 9 chairs ... there are 89 acceptable seating arrangements.

More generally you can get the acceptable arrangements for n chairs by seating a girl on the end of each $(n - 1)$ arrangement and a girl–boy on every $(n - 2)$ arrangement, giving us $F_n = F_{n-1} + F_{n-2}$ the equation that generates the Fibonacci numbers.

06

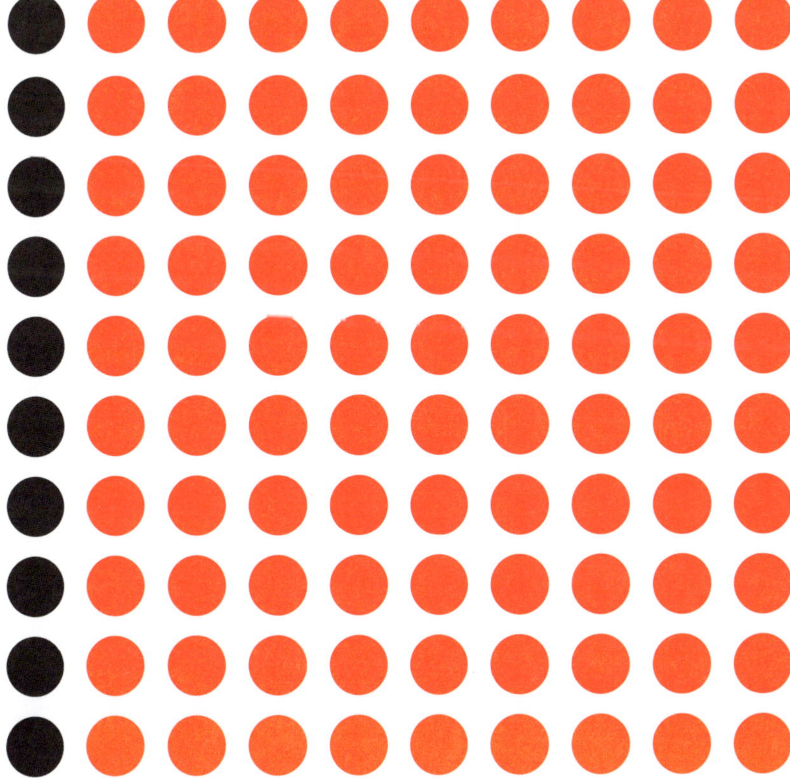

Ninety is an abundant number (see chapter 12) and there are 90 degrees in a right angle. In a square all internal angles are 90 degrees as are the external angles. Ninety is also a Harshad number (see chapter 84).

In medieval times, a 'moment' was officially defined as 90 seconds. We know this from a 1398 text in which the writer states that there are 40 moments in an hour.

Now, a bingo card usually has 90 numbers. Therefore there are approximately 44 million ways to make 'bingo'. Ah, Grandma, now I see why it's so exciting!

Each half of a standard top-flight football match is 45 minutes, so games go for 90 minutes. Except for games that go to extra time or golden goal extra time or penalty shoot-outs and, of course, allowing for injury time in both the first half and second half. In fact, the more I think about it, no games of football ever go for 90 minutes.

The pressure inside a champagne bottle is typically 90 pounds per square inch, according to the American Academy of Ophthalmology. This pressure can launch a 30-gram champagne cork at speeds of 80 kilometres per hour (and up to 13 metres in the air).

You and every person have nearly 400,000 radioactive atoms disintegrating into other atoms in your body each second. But there's no need to worry about falling apart. Each body cell contains an average of 90 trillion atoms – 225 million times that 400,000.

A 270° TRIANGLE?

We all know that the angles of a triangle add up to 180 degrees. Strictly speaking, that's not always true. It is whenever the triangle is drawn on a flat surface like a table or a blackboard. But when we draw triangles on curved surfaces, like a sphere or a doughnut, the 180-degree rule goes out the window.

Imagine you were at point A on the equator and travelled along the equator one-quarter of the Earth's circumference to point B, headed north to the North Pole C and then turned 90 degrees to your left and headed back to the equator at A.

The triangle ABC that you would have traced out on the curved surface of the Earth has three angles, each of 90 degrees making a total of 270 degrees.

PERRIN NUMBERS

Ninety is a Perrin number. Perrin numbers are defined by what mathematicians call a 'recurrence relation'. We've met these in chapters 70 and 86. It's just a rule where you do things to numbers in the series to get the next number in the series. The rule that defines the Perrin numbers is:

$P(0) = 3, P(1) = 0, P(2) = 2$ and

$P(n) = P(n - 2) + P(n - 3)$ (for $n > 2$)

So when $n = 3$ we get

$P(3) = P(1) + P(0) = 0 + 3 = 3$,

$n = 4$ gives us

$P(4) = P(2) + P(1) = 2 + 0 = 2$

... and so on.

QUIZ QUESTION: **Prove P(16) = 90 that 90 is the 16th Perrin number.**

ANSWER AT THE BACK OF THE BOOK

The world's
fastest
hummingbird
can flap
its wings
90 times
per second.

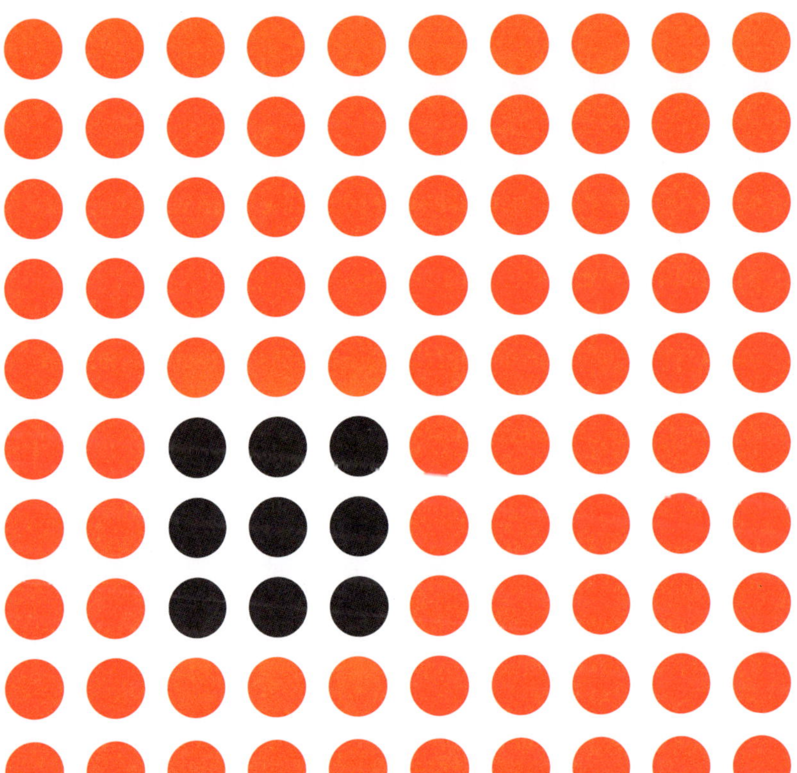

91

LITTLE THEOREM, BIG DEAL

We're getting pretty close to the end so let's look at something that is a little bit heavy and may take some rereading when we're finished for you to fully get it, but is also one of the most famous theorems in a branch of mathematics called Number Theory.

The number 91 is what's known as a 'pseud'. By that, I don't mean it's pretentious or superficial – no, it's a 'pseudoprime'. A pseudoprime is a number that 'looks' prime but isn't. That just means it passes some of the major tests for being a prime but is actually a composite number.

One of the rules that primes obey was discovered by the great mathematician Pierre de Fermat who we've already met many times in this book.

Fermat showed that if you have a prime number p and a number a such that a is not divisible by p then:

p is a factor of the number $a^{p-1} - 1$.

That sounds pretty hardcore but it's really not that difficult to understand. Here's an example:

7 is prime and 7 does not divide into 10, so 7 should be a factor of $10^6 - 1$ and it is: $10^6 - 1 = 1{,}000{,}000 - 1 = 999{,}999 = 7 \times 142{,}857$.

Similarly, 5 is prime and 5 doesn't divide into 2, so according to Fermat's 'Little' Theorem, as we call it:

5 should divide into $2^4 - 1 = 16 - 1 = 15 = 5 \times 3$

But while all primes obey Fermat's theorem, just because a number obeys the theorem for some choice of a, or even every choice of a, doesn't automatically make it prime.

Thus, when we take $n = 91$ and $a = 3$, we see that 91 is not a factor of 3 and 91 is a factor of $3^{90} - 1$. Trust me on that one.

So 91 looks prime when we use $a = 3$ but it clearly isn't. $91 = 7 \times 13$. So we call it a Fermat pseudoprime.

Other Fermat pseudoprimes include $n = 15$ which looks prime when you take $a = 4$:

$$a^{n-1} - 1 = 4^{14} - 1 = 268{,}435{,}455 = 15 \times 17{,}895{,}697$$

... but is clearly not prime because $15 = 5 \times 3$

And 4 when you take $a = 5$ and get $5^3 - 1 = 124 = 4 \times 31$

And 9 when you take $a = 8$ and get $8^8 - 1 = 16{,}777{,}215 = 9 \times 1{,}864{,}135$

... and so on and so on. Like I said, that's pretty heavy and normally not something you'd come across until university mathematics so if your head is hurting just a little bit that's understandable. Read it over again and it should sink in.

SHINE ON

It produces enough energy in one second to power the world for hundreds of millions of years. Light released from its core takes 100,000 years to reach the surface and then just 8 minutes to reach Earth. And it's estimated that it will continue to produce energy for another 7 billion years. You guessed it: I'm talking about the Sun.

The Sun is almost entirely made up of hydrogen and helium, but the ratio depends on how you calculate it. When it comes to the total number of atoms, hydrogen makes up about 91%, helium almost 9% and 65 other elements are present in small amounts.

But when you consider that the mass of an atom of helium is about 4 times that of an atom of hydrogen, the Sun can be considered 71% hydrogen, 27% helium and 2% other stuff.

TRIANGULAR

Ninety-one is triangular, square pyramidal and centred hexagonal. Trifecta!

QUARTER YEAR

Ninety-one is the number of days in a quarter year: 13 weeks 7 days each.

TENNIS

If, like me, you're wondering why you're no good at tennis, it might be because the middle of a tennis net is 91 centimetres high.

Or it might have something to do with the racquet, the balls or the alignment of the planets.

ABACI

A standard Chinese abacus – or *suanpan* – has 7 beads on each rod, two on the top deck and five below:

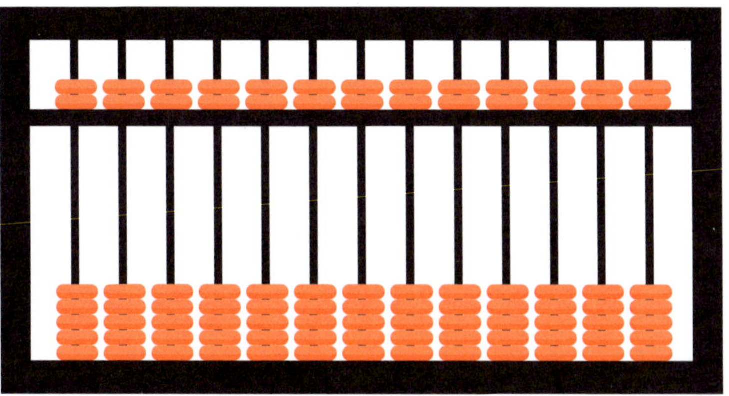

Such a *suanpan* with 13 rods has a total of 91 beads. Incidentally, they've been in use in China since at least the second century BC.

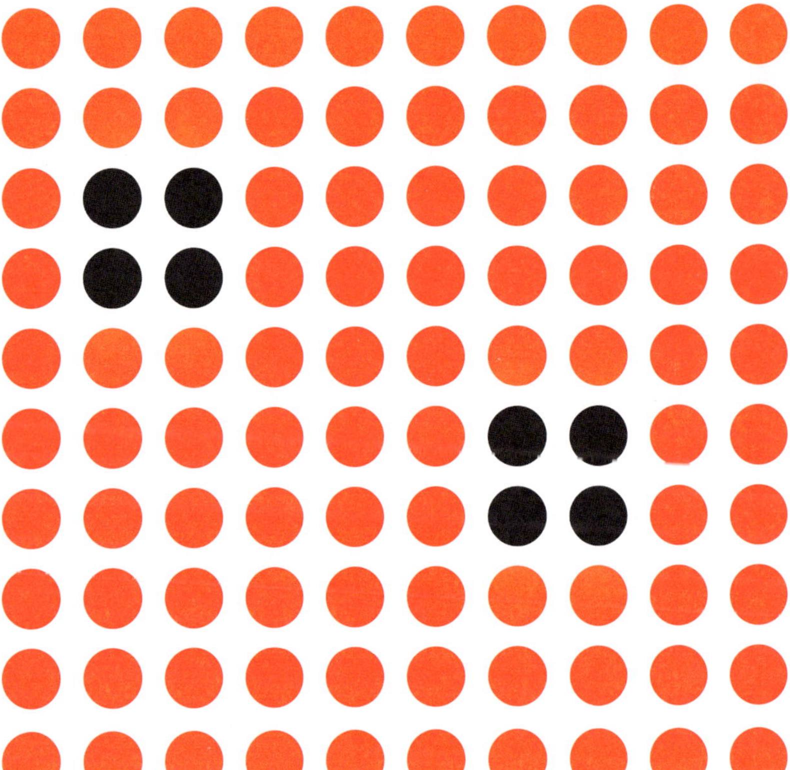

92

92 QUEENS

In chapter 12 we met the queen, the über-powerful chess piece that can attack all the squares along the row and column (in chess we say 'rank and file') and diagonals upon which she sits.

I challenged you to find the 12 ways you could place 8 queens 'safely' on a chessboard, that is, so that none of them could capture each other. But notice that in the question, I specified 12 arrangements that were different even when you rotated and reflected them.

If you count rotations and reflections of a solution to be new solutions, there's a lot more than 12 ways to place 8 queens safely on the board.

For example, these 8 arrangements:

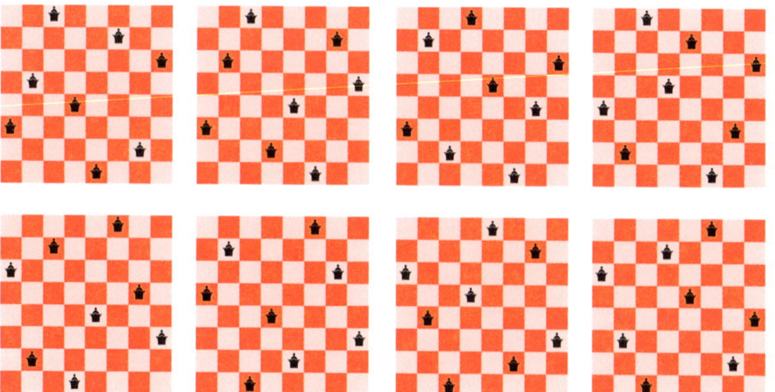

... are all just rotations and reflections of 1 of the original 12 solutions.

You might think, 'Okay, so all 12 solutions can be rotated and reflected 8 ways, so we get 12 × 8 = 96 general solutions'. But actually you only get 92.

That's because this solution:

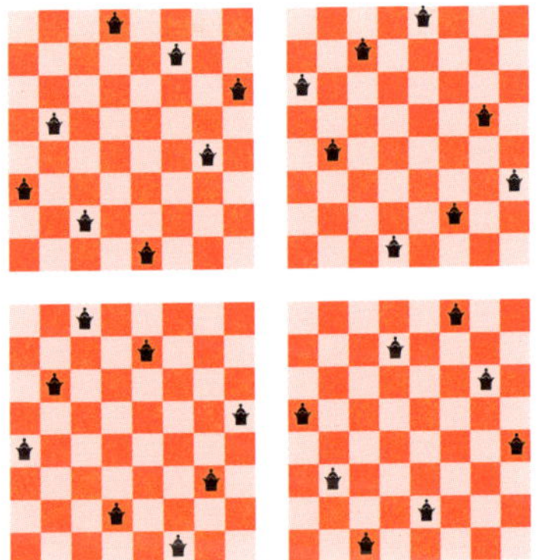

... does not give 8 different positions when rotated and reflected – only 4.

So there are 8 × 11 + 4 = 92 solutions to the queens problem when we allow rotations and reflections.

As we noted in chapter 55, 92 is one of the 28 numbers from 1 to 100 that can't be written as the sum of distinct squares.

A 'SNUB DODECAHEDRON'

... has 92 faces, 60 vertices and 150 edges. Check that Euler's formula $V + F - E = 2$ applies to this Archimedean solid.

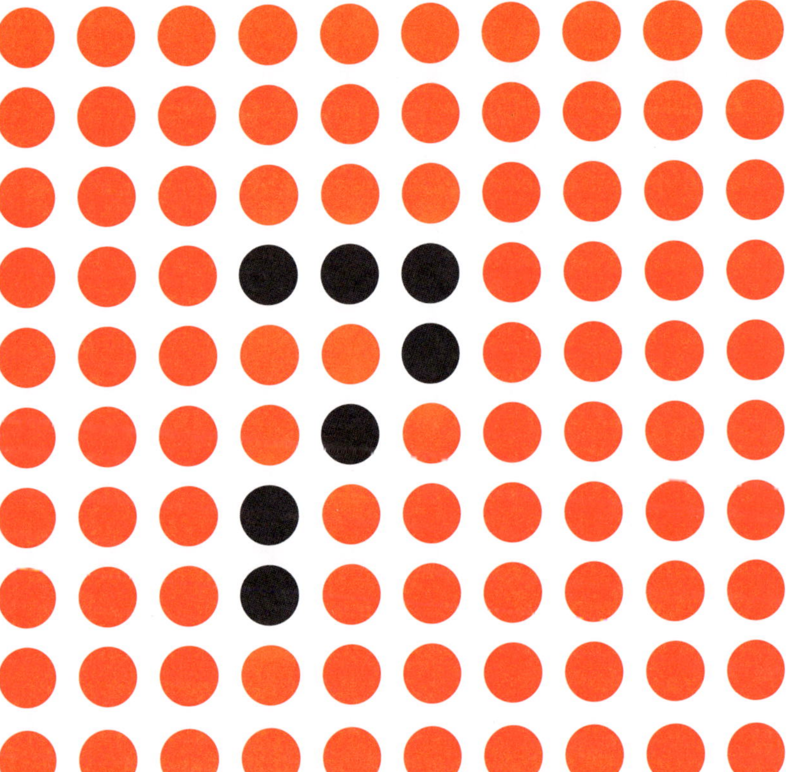

93

Ninety-three is the 22nd lucky number and back in chapter 25 I asked you to find all said lucky numbers up to 100.

In the next chapter we'll hear about Goldbach's conjecture, a famous piece of mathematics. Well, we have a similar but much less famous conjecture for the lucky numbers, namely that every even number can be written as the sum of two lucky numbers.

This has been proven for all even numbers up to well above 100,000 but I don't expect you to go that far. Perhaps you'd like to check it for all even numbers up to 100?

ASTRONOMICAL DISTANCE

The Sun is, on average, 93 million miles from Earth (about 150 million kilometres). So 93,000,000 miles, or an 'Astronomical unit' (AU) is one of the basic distances used for measuring in cosmology.

Let's try and get our heads around how big some things in our Universe are in terms of AU. These distances are only averages of course because objects are orbiting around stars and their orbits aren't even perfect circles and all sorts of other complicated stuff, but they give you a rough guide as to how incredibly far away some things are:

The distance from the Sun to
Earth is 1 AU
Mercury 0.4 AU
Jupiter 5.5 AU
Neptune 30 AU (about 4000 years drive)

The 'Termination Shock' where the solar wind from the Sun (travelling at 400 kilometres a second!) hits the rest of space (the interstellar medium) occurs about 80 to 200 AU from the Sun.

1 light year is 63,240 AU

The limit of the Sun's gravitational dominance is said to be the Oort Cloud, a theoretical cloud of icy objects – 100,000 AU from here.

The nearest star to our Sun (actually it's a group of three stars called Alpha Centauri A, B and Proxima Centauri) is located 277,600 AU away.

The distance across the Milky Way galaxy: 6,329,671,700 AU.

So if we scaled the Universe so that the Earth was 1 centimetre from the Sun (I know we'd be very hot, but go with me on this), the distance across the Milky Way galaxy would be 1 and a half laps of our regular-sized Earth.

But of course the Milky Way isn't the only galaxy in the Universe – it *is* beautiful with its spiral shape, probably a massive black hole at the centre, hundreds of billions of stars ... and of course us! – but we guess that in the Universe that we can see, there are about 170 billion (170,000,000,000) galaxies.

And speaking of 93, the distance across the observable Universe is about 93 billion light years or 5,881,320,000,000,000 (almost 6 quadrillion) AU.

MORE SEMIPRIMES

Ninety-three is semiprime because when written as the product of prime factors we only get two factors, namely $93 = 3 \times 31$. Chapter 86 also told us that 85, 86, 87 formed the second triplet of consecutive semiprimes.

Well, 93, 94, 95 is the third:

$93 = 3 \times 31$; $94 = 2 \times 47$; $95 = 5 \times 19$

QUIZ QUESTION: **Find all 34 semiprimes between 1 and 100.**

ANSWER AT THE BACK OF THE BOOK

DUDE, DON'T 93 ME!

According to Jonathan Bloom, Professor of Islamic and Asian Art as Boston College in the US, Persian poets would write of someone being greedy by saying that their hand 'made 93'.

It comes back to the practice of 'Dactylonomy' or representing numbers by holding your fingers in certain patterns. The 'arithmetic of the knots' as it was called by some was widespread in the ancient Arab World and medieval Europe and in the Persian case the number 93 was represented by a closed fist. So 'making 93' meant to offer someone a closed hand, or nothing.

EVERY DAY I'M SHUFFLIN'

Take 11 cards and deal as many as you want off the top of the deck. This gives you a new pile of cards with the card that started at the top of the original deck now at the bottom of the new deck.

Now, riffle the cards (or perhaps I should say, 'look up the definition of "riffle" on Wikipedia *then* riffle the cards'). Don't worry, riffling is just the fancy move you've seen people do when they have two piles of cards, raise the corners of each pile slightly, bring the piles really close and then 'riffle' them past their thumbs so the corners fall over each other and they can knock them together back into one pile.

Right. So riffle the cards.

You have now shuffled your original pile of 11 cards. In fact, when you shuffle this way, there are 93 × 11 different ways the new deck can come out – or 93 ways if you consider cycles of cards to be the same, for example, 3, 4, 5, 6, 7 the same as 5, 6, 7, 3, 4.

Such a shuffle is called a 'Gilbreath Shuffle'. It's slightly different to the usual riffle where you just split the deck because forming the second pile this way reverses the order of the cards you take off the top.

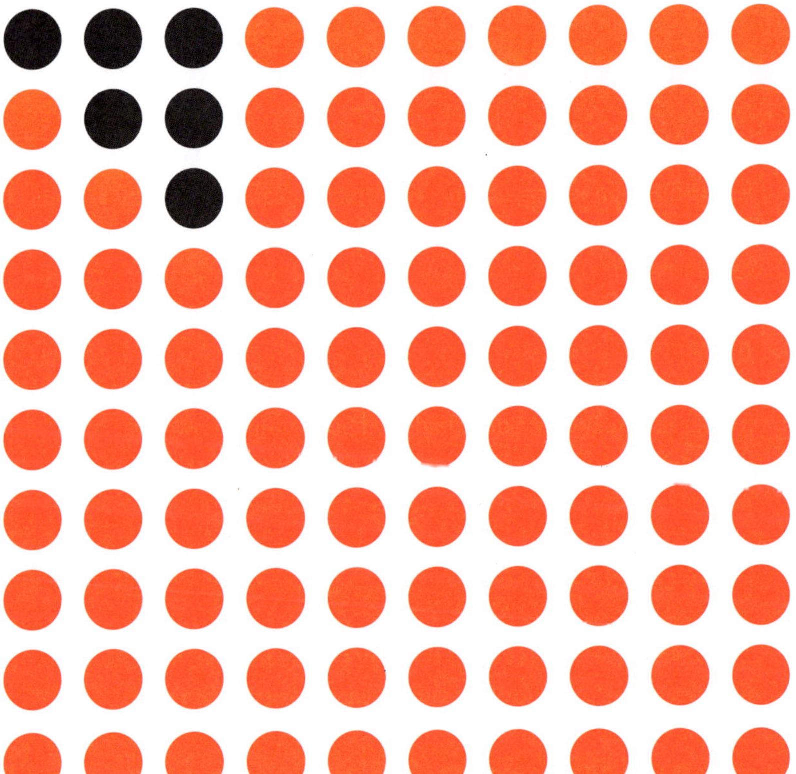

94

SMITH NUMBERS

Mathematicians are a very special bunch. Some are so beautifully in tune with the numerical world they see patterns *everywhere* – places most people never would. Ramunujan realised that 1729 was equal to both $1^3 + 12^3$ and $9^3 + 10^3$ when the famous mathematician G. H. Hardy mentioned it as the number of a taxi cab.

Another such numerical noticer was Albert Wilansky whose brother-in-law Harold Smith had the phone number 4937775.

So what, you may ask? Well, Albert noticed that when you factorise 4937775 into primes you get:

$$4937775 = 3 \times 5 \times 5 \times 65837$$

... and when you add together the digits of the number and the digits of its prime factorisation you get the same result, that is:

$$4 + 9 + 3 + 7 + 7 + 7 + 5 = 42$$
$$\text{and } 3 + 5 + 5 + 6 + 5 + 8 + 3 + 7 = 42$$

This is what defines the Smith numbers!

So to see that 94 is a Smith number, we first factorise it into primes – $94 = 2 \times 47$ – and then compare the sum of the digits of the number and its prime factorisation:

$$9 + 4 = 13 \text{ and } 2 + 4 + 7 = 13$$

So 94 is a Smith number.

QUIZ QUESTION: By definition, Smith numbers are always composite. Apart from 94, find the other 5 Smith numbers between 1 and 100.

HINT: One is less than 10, two are in the 20s and the other two are reflections of each other, for example, 37, 73. ANSWER AT THE BACK OF THE BOOK

THE HANCOCK HUSTLE

On the last Sunday of every February participants in 'Hustle up the Hancock' absolutely flog themselves up the 94 floors of Chicago's John Hancock Center to raise millions of dollars for respiratory health.

In 2014, Eric Leningar finished the climb in just 9 minutes 42 seconds, the fastest of all 2661 runners. Women's record holder Cindy Harris was the fastest of the 1261 female hustlers in a shade under 12 minutes.

As a proud Australian I should mention the feats of super stair scaler Suzy Walsham who has dominated the famous Empire State Building winning the 86-storey, 1576-step race on no fewer than 5 occasions! You go girl.

GOLDBACH'S CONJECTURE

In a letter to Euler on 7 June 1742, the German mathematician Christian Goldbach came up with what's known as Goldbach's conjecture. Like Fermat's Last Theorem, it's a great example of something that is very easy to explain but has been devilishly hard to prove.

Goldbach's conjecture is that 'every even number greater than 2 can be written as the sum of two prime numbers'.

For example, $94 = 5 + 89 = 11 + 83 = 23 + 71 = 43 + 51 = 47 + 47$.

As of 2012, Goldbach's conjecture had been shown to hold for all even numbers up to 4,000,000,000,000,000,000.

But again, this is not a proof for all numbers. Gee we mathematicians can be demanding, can't we? It's the rules of the game, sorry.

Why not prove that Goldbach's conjecture holds for all even numbers from 4 to 100?

Another idea, also contained in his letter to Euler and known as Goldbach's 'weak' conjecture is that 'every odd number greater than 5 can be written as the sum of three primes'.

Again, this seemed elusive until the Peruvian mathematician Harald Helfgott proved it to be so in 2013.

Background: Letter from Christian Goldbach to Leonhard Euler (1742). Source: Public Domain

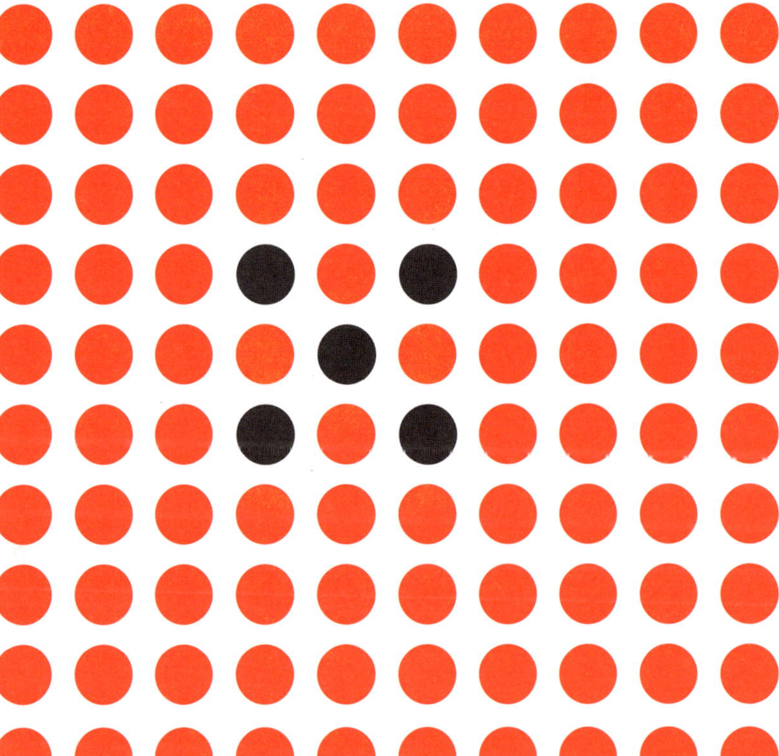

95

In our modern world we regularly interact with incredible machines and devices that most of us take for granted. Read one day about how an mp3 player works (or a CD machine if you're old school!) and you'll never look at one the same way again.

Take the humble barcode. While barcodes may look like a bunch of black lines in a box, they are actually a beautiful fusion of mathematics and technology. It's too small for us to see with our naked eyes, but a barcode is actually 95 separate columns of equal width that tell the scanner everything it needs to know about the product being scanned.

A laser scans the barcodes and the black bits are recorded as ones and the white bits as zeros. The ones and zeros of the barcode tell the scanner a 12-digit number – which is also printed under the barcode in case the scanner can't read it and an old-fashioned human is needed to type it in.

The 12-digit number on a packet of Kimberley Clark Huggies Baby Wipes tells you:

Product type (0 groceries)

Manufacturer's number (36000 Kimberley Clark)

Product number (29145 Huggies Baby Wipes)

Check digit (2)

Note the 'separation bars' so the scanner can see where numbers begin and end. The first and last digit sit in the 'quiet zone' outside the lines of information to tell the scanner where the barcode begins and ends

What's too small to see with the naked eye is that each digit of the barcode is really a set of 6 lines that are either white or black – this is why some lines look thicker than others. This tells the machine what each of the numbers are. For example:

(space, bar, space, bar, bar, bar) is the pattern for the number 6 on a barcode.

There are other really neat things about the way barcodes are designed – the small lines on the right-hand side are actually written in reverse to the left-hand side (zeros become ones and vice versa) so if you swipe the item upside down you don't confuse the machine into thinking that a $3.50 bag of apples is actually a $3000 flat screen TV.

The 'check digit' is the very cool way that the scanner 'checks' that it hasn't made a mistake. Every time you swipe an item, the scanner instantly calculates a number that is equal to:

(all the digits in the odd number positions) × 3 + (all the digits in the even positions)

The check digit is given by 10 minus the last digit of this number.

So for our Baby Wipes, we get:

$$(0 + 6 + 0 + 2 + 1 + 5) \times 3 + (3 + 0 + 0 + 9 + 4) = 42 + 16 = 58$$

The last digit of that number is 8, so the check digit is 10 – 8 = 2.

If the scanner has read the information as having a different check digit to what is on the barcode, the device beeps and the lovely person at the checkout types the barcode number in.

So next time you're getting grumpy at the shops because the machine won't read the barcode, remind yourself that you might just have saved $2996.50!

QUIZ QUESTION: Calculate the check digit on the barcode of this book. Note, it may be written to the left of the barcode. ANSWER NOT(!!!) AT THE BACK OF THE BOOK (you should know if you've got it correct)

Ninety-five is a number that appears in several Pixar films including *Toy Story*, *Toy Story 2*, *Toy Story 3*, *A Bug's Life*, *Cars* and *Cars 2*. It is a reference to the year 1995 when Pixar's first movie *Toy Story* was released.

SATURATING SATURN

The atmosphere on Mars is 95% carbon dioxide with a little bit of nitrogen, argon, oxygen and carbon monoxide thrown in. This unfriendly mix is one of the major obstacles to any planned human exploration or settlement on the red planet. That and what seems to be a complete lack of cafés and entertainment venues.

What about Saturn? Well, Saturn is 95 times heavier than the Earth and about 10 times as wide. You could fit over 750 Earths inside Saturn. But while it is massive, Saturn is a gas giant and the least dense planet in the solar system. So gravity on Saturn is about the same as gravity on Earth – its greater size and lesser density effectively cancel each other out.

But before you're thinking of taking a walk on Saturn I should point out that being a gas giant it doesn't have a solid surface.

The other amazing thing about the composition of Saturn is that it is so light if you had a big enough body of water, it would float on it!

96

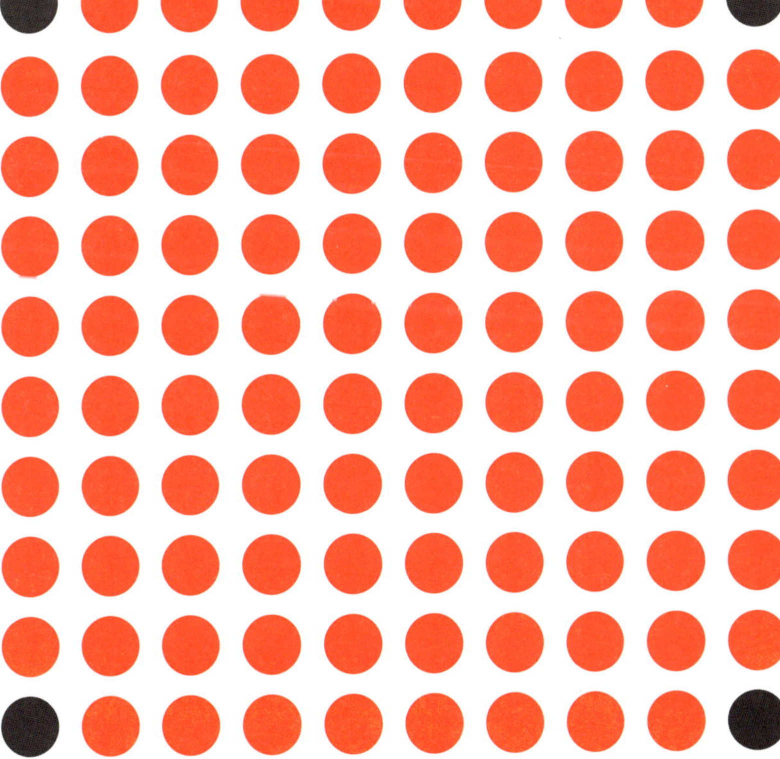

MEASUREMENT OF THE CIRCLE

In *Measurement of the Circle*, the Greek mathematician Archimedes showed that the exact value of π lies between the values $3\frac{10}{71}$ and $3\frac{1}{7}$. He worked this out by calculating how tightly a circle would fit inside two regular polygons that each had 96 sides. The great Chinese mathematician Liu Hui used the same shapes in the third century to get even more accurate results. Eureka!

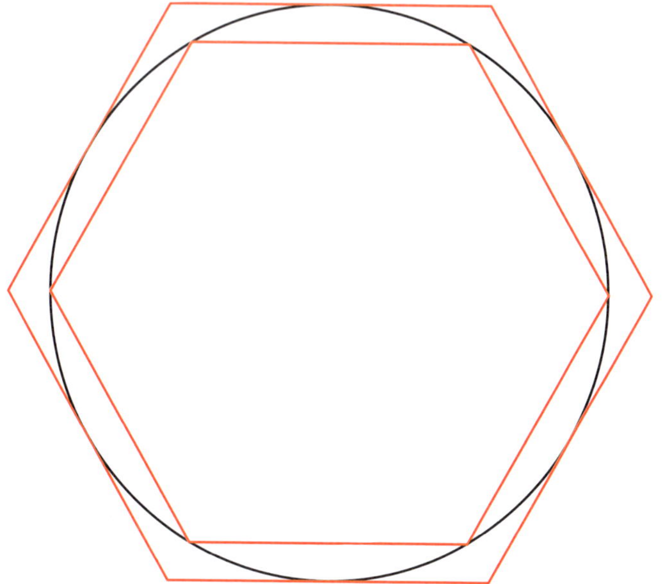

Put simply, Archimedes and Liu both started with the observation that the distance around the circle is in between the distances around the smaller and larger hexagons. They then kept doubling the sides of the shapes to make the inside and outside distances get closer and closer to the perimeter of the circle.

They didn't need to draw the 12-sided figure (which we can just call a 12-gon) or the 24-gon, 48-gon and 96-gon; they just kept calculating the perimeters until they produced their amazing results.

ABUNDANT

Ninety-six is the last abundant number before 100 itself which is also abundant.

UNTOUCHABLE

Ninety-six is also an untouchable number (see chapter 52).

RACY

The airing of the racy 1970s TV show *Number 96* is often said to be the moment Australia 'lost its innocence'.

OCTAGONAL NUMBERS

Ninety-six is an 'octagonal' number – which does not mean it's connected to the great Australian racehorse that won the 1995 Cox Plate and was named the 1996 Australian racehorse of the year.

In fact, I'm not going to tell you what an octagonal number is. I want you to look at the definitions of triangular and pentagonal numbers (chapters 21 and 22) and work it out for yourself. You'll know you're right if you can show 96 is an octagonal number.

QUIZ QUESTION: **We're almost at the end of our odyssey so why don't you go back and find all 22 of the abundant numbers from 1 to 100?** A few hints for you: 1) The abundant numbers from 1 to 100 are all even (the first odd abundant number is 945). 2) Every multiple of an abundant number is abundant. 3) Every multiple of a perfect number (except the perfect number itself) is also abundant, so look them up (chapters 6 and 28). Off you go!

ANSWER AT THE BACK OF THE BOOK

Teams with numbers in their names feature in sports the world over.

In Australia we have basketballers the Adelaide 36ers and the Sydney Sixers Cricket team to name two. America has the NBA Philadelphia 76ers and NFL San Francisco 49ers among many others.

But when it comes to numbered team names you can't go past the various leagues of German football. Teams are replete with names usually referring to the year in which they were formed or merged with other teams. You can choose from TSG 1899 Hoffenheim, FC 08 Homburg, Schalke 04, FC Hanau 93, FC Altona 93, SV Babelsberg 03, Bischofswerdaer FV 08, SV Darmstadt 98, SV Dessau 05, TSV 1860 München, Goslarer SC 08, Göttingen 05 and the comparatively recent FC Gütersloh 2000.

Two of my favourites are a club from the 7th flight of German football (the Landesliga Berlin Staffel 2), the mighty Berliner FC Alemannia 1890 and from the elite Bundesliga Hannover 96, which bears a logo that both looks pretty cool and shows how 96 reads the same (symmetrically) when turned upside down.

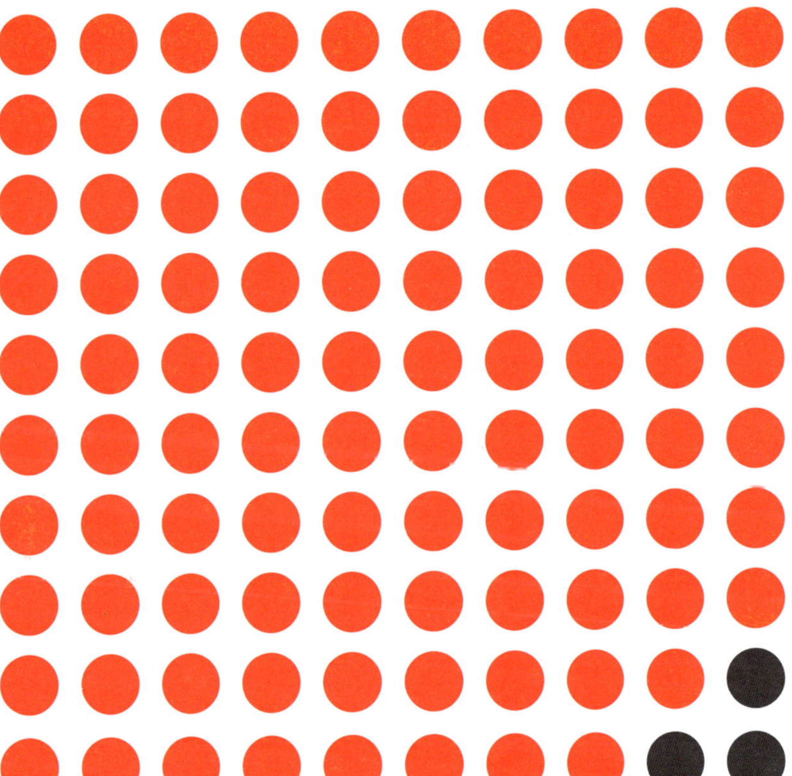

97

RECIPROCALS

We took a look at reciprocals and their decimal expansions way back in chapters 17 and 39. But that was a while ago so let's refresh our memories.

When you take a fraction and flip it over, you get its 'reciprocal'. So the reciprocal of $\frac{3}{4}$ is $\frac{4}{3}$.

The whole number 7 can be written as $\frac{7}{1}$ so its reciprocal is $\frac{1}{7}$. The reciprocal of 8 is $\frac{1}{8}$.

But while they look pretty similar, $\frac{1}{7}$ and $\frac{1}{8}$ are different in a very important way. When we write these fractions as decimals,

$$\frac{1}{8} = 0.125 \text{ but}$$

$$\frac{1}{7} = 0.142857\ 142857\ 142857\ 142857\ \ldots$$

The decimal expansion of 7 doesn't stop or 'terminate' as we mathematicians like to say. The numbers 142857 repeat forever and ever and are called the 'repetend' of $\frac{1}{7}$. So we say that $\frac{1}{7}$ has a repeating decimal expansion.

This is not the same as the decimal expansion like:

$\pi = 3.14159265358979323846264338327950288419716939937510582097494459 2\ \ldots$

or $\sqrt{2} = 1.4142135623730950488016887242096980785696\ \ldots$

... which both also go on forever but don't repeat. These types of numbers, with non-repeating, non-terminating decimal expansions, are called 'irrational' and can't be written as fractions. We can only choose fractions close to them, like $\frac{22}{7}$ for π and $\frac{41}{29}$ for √2.

For any prime number p, if the fraction $\frac{1}{p}$ has a repeating decimal expansion and the repetend is $(p - 1)$ numbers long, we call that repeating bit a 'cyclic' number.

So for $\frac{1}{7}$ the repeating section of the decimal, '142857' is 6 digits long and this makes 142857 a cyclic number.

A cyclic number has the cool property that as you multiply it, it 'cycles through itself' so:

$1 \times 142857 = 142857$
$2 \times 142857 = 285714$
$3 \times 142857 = 428571$
$4 \times 142857 = 571428$
$5 \times 142857 = 714285$
$6 \times 142857 = 857142$

Similarly, as we saw in chapter 17:

$$\frac{1}{17} = 0.0588235294117647\ 0588235294117647\ 0588235294117647 \ldots$$

So the 16-digit repetend 0588235294117647 is a cyclic number.

For the numbers 1 to 100, the reciprocals of 7, 17, 19, 23, 29, 47, 59, 61 and 97 all generate cyclic numbers. Note that all these numbers are prime and are called 'full repetend primes'.

The cyclic number generated by $\frac{1}{97}$ is this 96-digit beast:

010309278350515463917525773195876288659793814432989690721649484536082474226804123711340206185567

... which, as you multiply it by 2, 3, 4 and so on, cycles through its digits. How about you just take my word for that?

PROTH NUMBERS

The number $97 = 3 \times 2^5 + 1$ and as such is a Proth prime.

A Proth number, named after François – you guessed it – Proth, is a number of the form $a \times 2^b + 1$ where a is an odd positive integer and b a positive integer such that $2^b > a$.

If a Proth number is prime, we call it a Proth prime.

So $97 = 3 \times 2^5 + 1$ is a Proth prime as is $17 = 1 \times 2^4 + 1$.

QUIZ QUESTION: The remaining Proth primes under 100 are 3, 5, 13, 17 and 41 – prove they are indeed Proth primes.

ANSWER AT THE BACK OF THE BOOK

A SONNET

William Shakespeare's Sonnet XCVII (97) has a curious property:

How like a winter hath my absence been
From thee, the pleasure of the fleeting year!
What freezings have I felt, what dark days seen!
What old December's bareness every where!
And yet this time remov'd was summer's time,
The teeming autumn, big with rich increase,
Bearing the wanton burden of the prime,
Like widow'd wombs after their lord's decease:
Yet this abundant issue seem'd to me
But hope of orphans and unfather'd fruit;
For summer and his pleasures wait on thee,
And, thou away, the very birds are mute;
Or, if they sing, 'tis with so dull a cheer
That leaves look pale, dreading the winter's near.

The 7th word of the 7th line of the sonnet is 'prime' – and 97 is the largest 2-digit prime number!

86

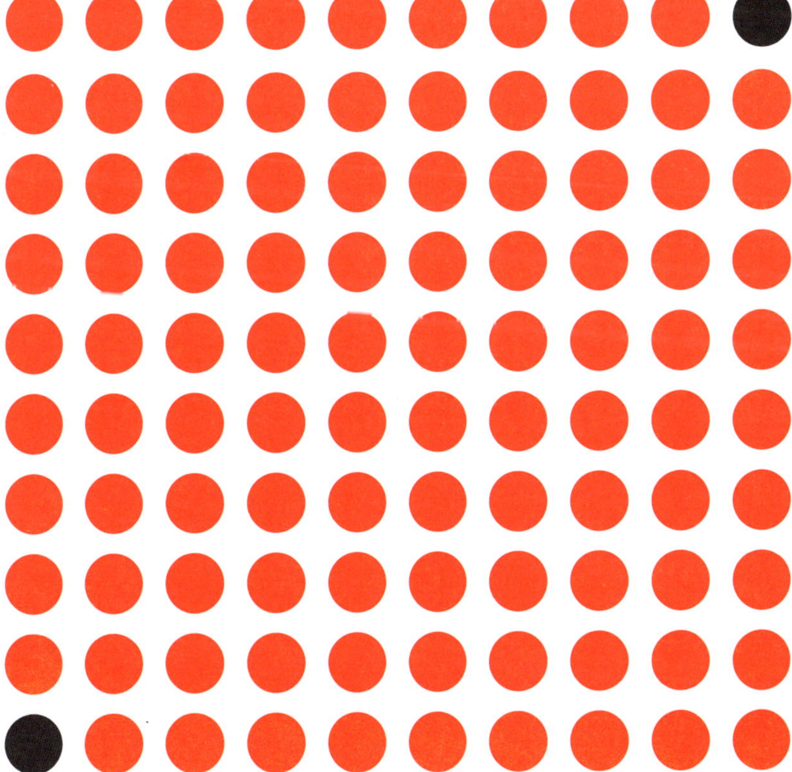

QUIZ QUESTION: A famous maths problem with a surprising answer is the 'potato problem'. You have 100 kilos of potatoes, which are 99% water. You leave them in the Sun and they dehydrate until they are 98% water. How much do they now weigh? ANSWER AT THE BACK OF THE BOOK

THERE ON GILLIGAN'S ISLE

Gilligan bumbled around, Ginger looked exceptionally well-presented for a castaway, and the Professor could build radios, weapons, houses and do just about everything other than FIX THE BOAT for 98 episodes of *Gilligan's Island*.

This doesn't include the TV spectacular *Baywatch Meets Gilligan's Island*, which must rank as one of the worst pieces of television I've ever been unfortunate enough to witness. And, believe me, I've seen some turkeys in my time.

NO PRIMES NEARBY

Goldbach's conjecture in chapter 94 suggested that all even numbers apart from 2 can be written as the sum of 2 primes. When it comes to adding odd primes together to get even numbers we see that $6 = 3 + 3$, $8 = 3 + 5$, $10 = 5 + 5 = 3 + 7$, $12 = 5 + 7$ ... and that most even numbers are 'close' to primes so we can find a sum that involves 3, 5 or 7 as one of the prime numbers.

Well, $98 = 3 + 95 = 5 + 93 = 7 + 91$ but 95, 93 and 91 are all composite numbers and 98 is the first number that when you write it as the sum of two primes needs a smaller prime greater than 7.

QUIZ QUESTION: Write 98 as the sum of two primes with one of those primes as small as possible. ANSWER AT THE BACK OF THE BOOK

URANUS' TILT

The most tilted planet in the solar system is Uranus, which rolls through space tilted at 98 degrees. As a result, summer on Uranus lasts up to 21 years, but for those of you calling your travel agent, please note that winter goes for just as long, during which overnight temperatures can get as low as minus 219° Celsius. Brrr.

CHIMPANME CHIMPANYOU

The often quoted statistic is that humans and chimps share about 98% of their DNA. At the same time, it's claimed on some websites that we share something like 92% with other mammals: 80% with certain cows and 40% with fruit flies. Some claim, genetically, you're about 25% similar to a grain of rice and 7% to some bacteria.

Catchy stats – but the reality is a lot more complicated. These numbers refer to the sequences of DNA that code for the proteins that are the building blocks of our bodies. The humble axolotl has some 20 times as much DNA as we do in every cell in their body, they have genes that we do not have, and they can regenerate their limbs (cool!). The important thing is the control network. We have pretty much the same body plan as a frog, and use most of the same parts to make it, but it looks and works rather differently!

So clearly it's what we do with our genes that can have a massive effect determining the difference between you and the average chimp. The genes in our brains and immune systems are the most variable between species because they are targeted by genetic selection, so if that's where there are significant differences you could argue that despite what some people call a very small "percentage DNA difference" we're not really that chimp-ish at all.

We still have along way to go to understand the complexity of our genetic code and what our DNA really means.

At the same time, how's the back hair going grandpa?

The normal human body temperature (measurements taken internally) is considered to be 98.6° Fahrenheit (37° Celsius).

BACHET'S CONJECTURE

In Bachet's famous 1612 book of 'delectable problems' (which also gives us Bachet's Problem, which we met in chapter 40) he said that it seemed to be the case that every positive number can be written as the sum of at most four squares.

So $16 = 4^2$ is the sum of just one square, $5 = 2^2 + 1^2$ only needs two squares. $38 = 5^2 + 3^2 + 2^2$ but can't be written as the sum of only one or two squares.

Now $71 = 6^2 + 5^2 + 3^2 + 1^2 = 7^2 + 3^2 + 3^2 + 2^2$

... so there are two different ways that 71 can be written as the sum of four squares – but turns out it can't be written using any less than four squares.

And Bachet was pretty sure that this was as gnarly as it gets. That any number can be written using at most four squares.

Thankfully in 1770 Lagrange put an end to sleepless nights on this one. By which stage Bachet was well asleep ... he died in 1638.

QUIZ QUESTION: 98 doesn't need four squares, it can be written as $98 = 7^2 + 7^2$ but find the two ways 98 can be written as the sum of three squares and the four ways it can be written as the sum of four squares. ANSWER AT THE BACK OF THE BOOK

It turns out that the only numbers for which you need all four squares are all the numbers of the form $n = 4^m(8k + 7)$, for m and k positive whole numbers or 0. So for example, when $m = 1$, $k = 0$ we get $n = 4 \times 7 = 28$ which needs all four squares.

66

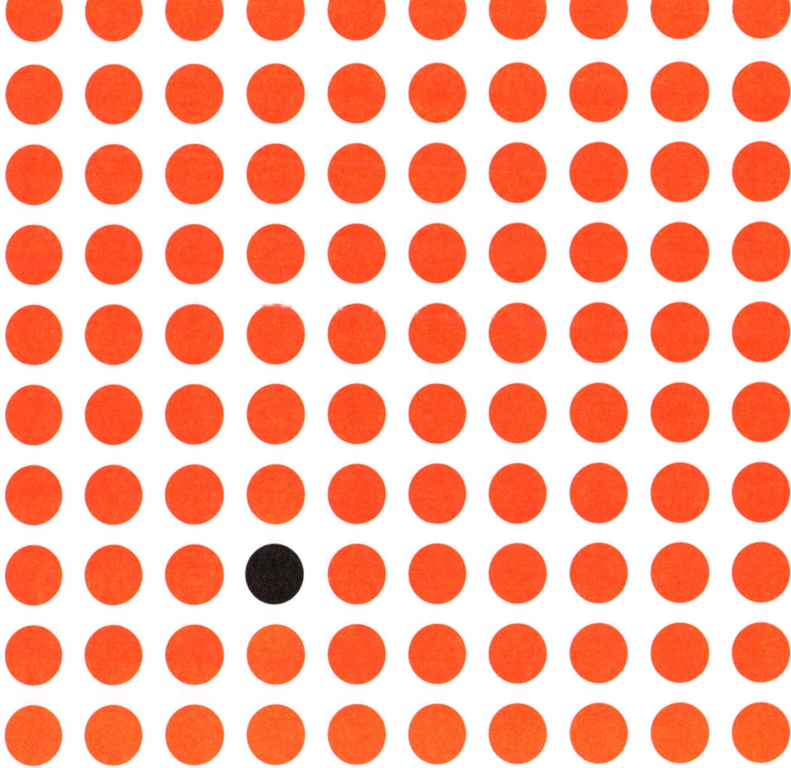

FRACT-TASTIC

We're close to the end of the book now so let's check out something that looks really nasty but isn't actually so bad. It's called a 'continued fraction'. Here's one I prepared earlier:

$$1+\cfrac{1}{1+\cfrac{1}{2+\cfrac{1}{1+\cfrac{1}{3}}}}$$

Working from the bottom up, remembering that to add fractions you put them all over the same denominator and that the reciprocal of $\frac{a}{b}$ $= \frac{b}{a}$ we get:

$$1+\cfrac{1}{1+\cfrac{1}{2+\cfrac{1}{1+\cfrac{1}{3}}}} = 1+\cfrac{1}{1+\cfrac{1}{2+\cfrac{1}{\frac{4}{3}}}} = 1+\cfrac{1}{1+\cfrac{1}{2+\frac{3}{4}}} = 1+\cfrac{1}{1+\cfrac{1}{\frac{11}{4}}}$$

$$= 1+\cfrac{1}{1+\frac{4}{11}} = 1+\cfrac{1}{\frac{15}{11}} = 1+\frac{11}{15} = \frac{26}{15}$$

So that monster we started with is the continued fraction of $\frac{26}{15}$. It is equal to $\frac{26}{15}$, it just looks different when written in this continued fraction form.

At this stage I should apologise to all high school maths teachers for whom these equal signs should be going down the page not across – I know it looks truly repugnant but it's getting near the end of the book and I'm a bit pressed for space. Sorry ... and I think yours is a truly noble occupation.

Anyway, all fractions can be written as continued fractions that stop after some number of steps. But we can also use continued fractions to approximate numbers that can't be written as exact fractions (that is, the irrational numbers) like:

$$\sqrt{2} = 1 + \cfrac{1}{2 + \cfrac{1}{2 + \cfrac{1}{2 + \cfrac{1}{2 + \dots}}}}$$

And one of my favourites:

$$\sqrt{99} = 9 + \cfrac{1}{1 + \cfrac{1}{18 + \cfrac{1}{1 + \cfrac{1}{18 + \cfrac{1}{1 + \dots}}}}}$$

The oldest person ever to file for divorce was 99-year-old Italian Antonio C who divorced his wife of 77 years when he uncovered love letters indicating an affair she'd had 70 years earlier while he was fighting in World War II.

NEUNUNDNEUNZIG AWESOME SONG

In June 1982, as the Rolling Stones released balloons into the air while playing a gig in West Berlin, little did they realise that Carlo Karges – guitarist for German pop outfit Nena – was in the audience. Watching the flock of balloons changing shape as they blew through the air, Carlos wondered what would happen if they drifted into East German airspace and were misinterpreted as a fighter jet or a UFO?

Thankfully, there was no East German nuclear response – but the song Carlos wrote about that exact scenario, '99 Luftballons', became a worldwide hit the next year and 30 years later is still responsible for some of the worst dancing by badly-dressed 45-year-old men you're ever likely to see.

QUIZ QUESTION: Prove that $\frac{1}{99} = 0.01010101010101\ldots$

HINT: See chapter 81. ANSWER AT THE BACK OF THE BOOK

99 BOTTLES OF BEER ON THE WALL

... 99 bottles of beer, if one of those bottles should happen to fall there'd be 98 bottles of beer on the wall; 98 bottles of beer on the wall 98 bottles of beer, if one of those bottles should happen to fall there'd be 97 bottles of beer on the wall; 97 bottles of beer on the wall 97 bottles of beer, if one of those bottles should happen to fall there'd be 96 bottles of beer on the wall; 96 bottles of beer on the wall 96 bottles of beer; if one of those bottles should happen to fall there'd be 95 bottles of beer on the wall; 95 bottles of beer on the wall, 95 bottles of beer; if one of those bottles should happen to fall there'd be 94 bottles of beer on the wall; 94 bottles of beer on the wall, 94 bottles of beer; if one of those bottles should happen to fall there'd be 93 bottles of beer on the wall; 93 bottles of beer on the

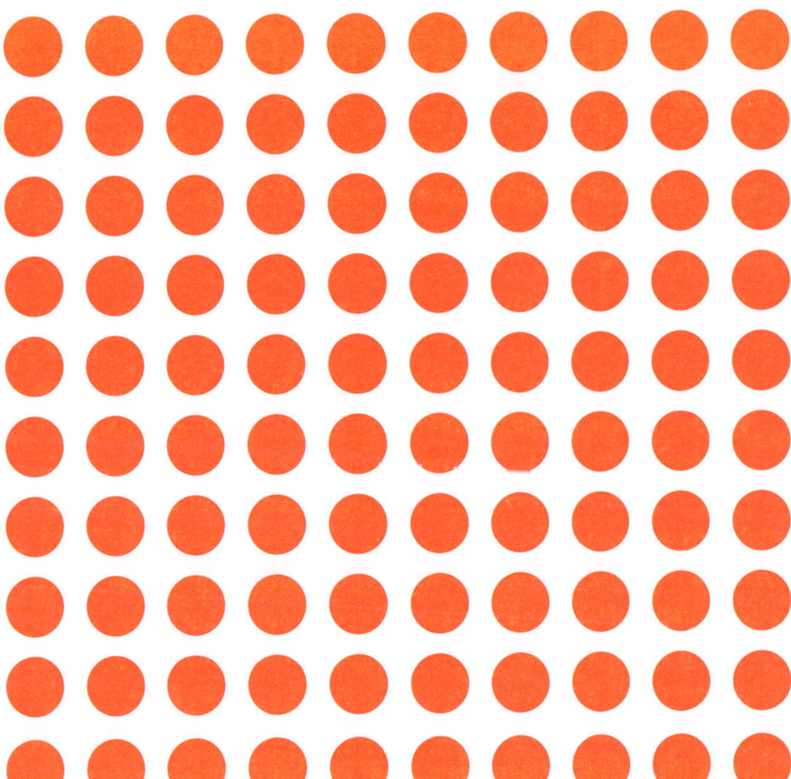

100

GOOGOLPLEXY AND I KNOW IT

We've met a lot of very big numbers in this book and here's a couple more. One of the most famous big numbers is a 'googol'. A googol is 10^{100} or one followed by 100 zeros:

10,000,000,000,000,000,000,000,000,000,000,000,000,000,000,000,000,000,000,000,000,000,000,000,000,000,000,000,000,000,000,000,000,000,000

The word was coined by 9-year-old Milton Sirotta when his uncle, the American mathematician Edward Kasner, asked for a name for 1 followed by 100 zeros. Strictly, it could also be called 'ten thousand sexdecillion' or 'ten duotrigintillion'. But I think we all agree 'googol' sounds much cooler.

A googol is about equal to $70! = 70 \times 69 \times 68 \times \ldots \times 3 \times 2 \times 1$ which is the number of ways that 70 objects can be arranged in order.

So if 70 school kids went to the museum, there are about a googol different ways they could line up to get in.

At the same time, the number of fundamental particles in the entire Universe is thought to be about 1 with 80 to 85 zeros after it.

This is worth thinking about – the number of ways 70 kids can line up at a museum door is the same as the number of fundamental particles in billions of billions of universes!

The tech behemoth Google is thought to actually be named after a misspelling of googol that the founders liked the look of.

So, I hear you ask, what is the name for the number 1 with a googol zeros after it?

It's called a googolplex. Fantastically large, it was preserved in pop culture by the brilliant writers of *The Simpsons* as the name of the Springfield Googolplex Theaters – the downtown cinema complex with 'at least 28 screens'.

In fact, *The Simpsons* is riddled with hilarious maths references as the great British author Simon Singh points out in his rollicking read *The Simpsons and Their Mathematical Secrets*.

Let's try and get our heads around how big these numbers actually are. According to my younger daughter Olivia, a googolplex kicks in any time a number gets bigger than about 20. In fact, it's a lot larger than that.

The American scientist Carl Sagan suggested it would not be possible to write out a googolplex 10,000,000,000 ... because you'd run out of space in the Universe. Physicist Don Page estimates that if you typed two zeros per second, you'd need about 10^{80} times the current life of the Universe to get there.

Estimates of the 'heat death' of the entire Universe – when everything has continued to expand and cool until what some people call 'the big freeze' kicks in – suggest in a googol years all black holes will have evaporated and the Universe will be practically empty. Great upbeat note to finish the book on, Adam!

TRIANGLES AND CUBES

There is a relationship between triangular numbers and cubes.

Take a triangular number and square it; it is equal to the sum of some consecutive cubes, starting at 1^3.

Put more formally, $(T_n)^2 = 1^3 + 2^3 + ... + n^3$.

Now, as I mentioned earlier, 10 is the fourth triangular number, because 10 = 1 + 2 + 3 + 4.

So in the formula above, $T_4 = 10$, and we can write:

$10^2 = 1^3 + 2^3 + 3^3 + 4^3$

or

$100 = 1^3 + 2^3 + 3^3 + 4^3$.

QUIZ QUESTION: Using only +, – or x, join the digits 1 to 9, in order, to make a total of 100. One way is 1 + 2 + 3 + 4 + 5 + 6 + 7 + (8 x 9) = 100. But there are others. Find another one. ANSWER AT THE BACK OF THE BOOK

GAUSS IS IN DA HOUSE

There's a story that does the rounds that the young German genius mathematician Carl Friedrich Gauss was at school one day when his teacher, J. G. Büttner, cracked it at the kids. As punishment, JGB set them a problem which he thought would keep the little tackers quiet for an hour or two. On the board he wrote:

$1 + 2 + 3 + \dots + 98 + 99 + 100 = ?$ and sat smugly at his desk.

According to legend, no sooner had his bütt hit the chair, than young CFG came forward with his chalk tablet covered in the correct answer. There's dispute as to whether this story is true, but I hope it is.

What Gauss had realised, and we saw in chapter 81, is that there is an easy way to add a long string of numbers like this, because it doesn't matter in which order you add them: $5 + 12 + 19 = 19 + 5 + 12$.

So you can rewrite $1 + 2 + 3 + \dots + 98 + 99 + 100$ as
$1 + 100 + 2 + 99 + 3 + 98 + \dots + 50 + 51$.

As long as you're careful that you've covered every number in the list you're fine doing this.

You can also group this new sum into pairs that all equal 101:

$$(1 + 100) + (2 + 99) + (3 + 98) + \dots + (50 + 51) = 101 + 101 + 101 + \dots + 101$$

And the 50 in the last bracket shows you you've got 50 brackets in the sum:

So $1 + 2 + \dots + 100 = 50 \times 101 = 5050$.

In your face, Mr Büttner!

Well, there you go: the numbers 1 to 100.

I hope you've enjoyed reading and learning about them as much as I enjoyed walking you through.

If there's anything that didn't make sense, please go back and have another look – these things often sink in a bit deeper on the second or third time round.

A marketing friend of mine suggested I tell you that if you purchase 10 copies of the book and sleep with them under your pillow, the knowledge will seep up into your brain at night. I don't believe this to be true, but if you want to try it ... don't let me stop you.

But, seriously, if you've got any suggestions, questions, or things you'd like me to know, my Twitter handle is @adambspencer, and there's a Facebook page as well as an interactive comments page for the book on my website at adamspencer.com.au

Time now for the answers to those questions that have been keeping you awake at night!

Quiz Answers

CHAPTER 3

Page 10: Cube; V=8, F=6, E=12; 8 + 6–12=2

Square pyramid; V=5, F=5, E=8; 5 + 5–8=2

Octahedron; V=6, F=8, E=12; 6 + 8 – 12=2

CHAPTER 4

Page 15: (1,1,1,1), (1,1,2), (1,3), (2,2), (4)

CHAPTER 9

Page 36 (1): 9567 +
1085

10,652

Page 36 (2): The number of letters in each word gives you 3141592653. And 3.141592653 is pi to 9 decimal places.

CHAPTER 11

Page 43: R3 = 111 = 3 x 37

R4 = 1111 = 11 x 101

R6 = 111,111 = 3 x 7 x 11 x 13 x 37

Page 44: You get a good idea when you try if you've got it right or not. The last few terms are 768; 336; 54; 20; 0.

CHAPTER 12

Page 46: A dozen, a gross and a score

Plus three times the square root of four

Divided by seven

Plus five times eleven

Is exactly nine squared, nothing more

Page 48: It's getting a bit crammed here in the answer section. Do me a favour and look online under 'queens problem'.

CHAPTER 17

Page 68 (1): 8, (8^3 = 512 and 5 + 1 + 2 = 8); 17; 18 (18^3 = 5832 and 5 + 8 + 3 + 2 = 18); 26, (26^3 = 17,576 and 1 + 7 + 5 + 7 + 6 = 26); 27 (27^3 = 19,683 and 1 + 9 + 6 + 8 + 3 = 27)

Page 68 (2): $17 = 3^2 + 2^3 = 4^2 + 1^3$

CHAPTER 18

Page 71: Amy and Ben cross (2) Amy returns (1); Delia and Cassius cross (10) Ben returns (2); Amy and Ben cross (2). This takes only 2 + 1 + 10 + 2 + 2 = 17 minutes. You save 2 minutes since, even though Ben has to make an extra 2 crossings compared to our 19-minute example, you save 5 minutes by sneaking Cassius across with Delia! Brilliant stuff!

CHAPTER 19

Page 76: F_{19} = 4181 = 37 x 113.

Ah, mathematics can be a cruel mistress.

CHAPTER 25

Page 98: 1, 3, 7, 9, 13, 15, 21, 25, 31, 33, 37, 43, 49, 51, 63, 67, 69, 73, 75, 79, 87, 93, 99

Page 99:

CHAPTER 27

Page 108: 26 and 28 should be fine – but I'm really impressed if you got 27, 82, 41, 124, 62, 31, 94, 47, 142, 71, 214, 107, 322, 161, 484, 242, 121, 364, 182, 91, 274, 137, 412, 206, 103, 310, 155, 466, 233, 700, 350, 175, 526, 263, 790, 395, 1186, 593, 1780, 890, 445, 1336, 668, 334, 167, 502, 251, 754, 377, 1132, 566, 283, 850, 425, 1276, 638, 319, 958, 479, 1438, 719, 2158, 1079, 3238, 1619, 4858, 2429, 7288, 3644, 1822, 911, 2734, 1367, 4102, 2051, 6154, 3077, 9232, 4616, 2308, 1154, 577, 1732, 866, 433, 1300, 650, 325, 976, 488, 244, 122, 61, 184, 92, 46, 23, 70, 35, 106, 53, 160, 80, 40, 20, 10, 5, 16, 8, 4, 2, 1

CHAPTER 28

Page 111 (1): $8128 = 1^3 + 3^3 + 5^3 + \ldots + 15^3$

Page 111 (2): Yes:

CHAPTER 29

Page 116: 20 and 21; $20^2 + 21^2 = 400 + 441 = 841 = 29^2$

CHAPTER 30

Page 119: 9 = 7 + 2 = 5 + 2 + 2 = 3 + 3 + 3 = 3 + 2 + 2 + 2

Page 120: 7# = 7 x 5# = 210

11# = 11 x 7# = 2310

13# = 30,030

CHAPTER 33

Page 130: 55 = 3 + 6 + 10 + 36, 64 = 3 + 6 + 55, 90 = 3 + 21 + 66

Page 131: It is a magic hexagram with constant 33.

CHAPTER 34

Page 134:

12	13	1	8
6	3	15	10
7	2	14	11
9	16	4	5

1	2	15	16
13	14	3	4
12	7	10	5
8	11	6	9

10	16	1	7
3	9	8	14
6	4	13	11
15	5	12	2

16	3	2	13
5	10	11	8
9	6	7	12
4	15	14	1

Note: in the bottom left square, we could swap the 3 with the 6 and the 14 with the 11 and have a new solution.

CHAPTER 36

Page 144: 1, 2, 4, 6, 12, 24, 36, 48, 60

CHAPTER 38

Page 150:

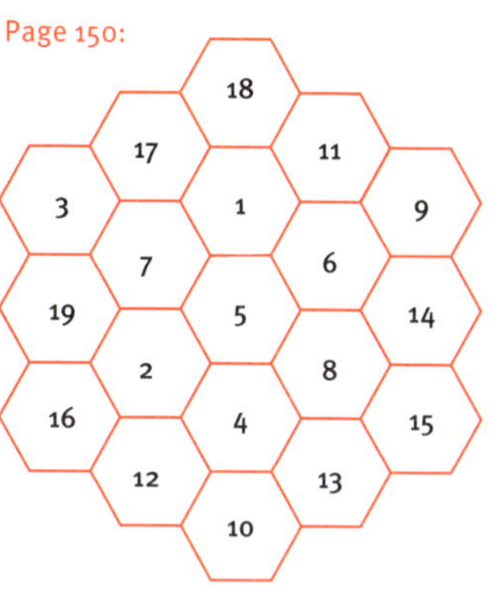

CHAPTER 39

Page 155 (1): (4,15,20), (5,10,24) and (6,8,25).

Page 155 (2): 407, 8208, 93,084, 1,741,725, 4,679,307,774

CHAPTER 43

Page 170: The twin pair primes between 1 and 100 are (3, 5), (5, 7), (11, 13), (17, 19), (29, 31), (41, 43), (59, 61) and (71, 73)

Page 172 (1): 1, 2, 3, 4, 5, 7, 8, 10, 11, 13, 14, 16, 17, 19, 22, 23, 25, 28, 31, 34, 37, and 43

Page 172 (2): 23, 27, 29, 17 and 89. More generally for every pair (a, b) you should get the Frobenius number $a \times b - (a + b)$

CHAPTER 45

Page 179: $45 = 1 + 2 + 3 + 4 + 5 + 6 + 7 + 8 + 9$, $45 = 7 + 8 + 9 + 10 + 11$, $45 = 14 + 15 + 16$. And strictly $45 = 45$ if you want a 6th way!

Page 180 (1): 55, 99, 297

Page 180 (2): $55^4 = 9{,}150{,}625$ and $9 + 15 + 06 + 25 = 55$ so the answer is 55

CHAPTER 47

Page 186 (1): 89, 179, 359, 719, 1439, 2879 is a Cunningham chain of length 6.

Page 186 (2): 1, 2, 3, 4, 6, 8, 11, 13, 16, 18, 26, 28, 36, 38, 47, 48, 53, 57, 62, 69, 72, 77, 82, 87, 97, 99

Page 188: 14, 19, 28, 47, 61, 75

CHAPTER 48

Page 191: 140 and 195 and 1575 and 1648

CHAPTER 49

Page 196: 444,444,888,889

CHAPTER 50

Page 198: $65 = 1^2 + 8^2$
$= 4^2 + 7^2$

CHAPTER 52

Page 208: (ABCD), (ABC)(D), (ABD)(C), (ACD)(B), (BCD)(A), (AB)(CD), (AC)(BD), (AD)(BC), (A)(B)(CD), (A)(C)(BD), (A)(D)(BC),(B)(C)(AD), (B)(D)(AC), (C)(D)(AB), and (A)(B)(C)(D)

CHAPTER 53

Page 212 (1): 89

Page 212 (2): The sum of the first 23 primes is $874 = 23 \times 38$.

CHAPTER 54

Page 216: $54 = 7^2 + 2^2 + 1^2 = 6^2 + 3^2 + 3^2 = 5^2 + 5^2 + 2^2$

CHAPTER 55

Page 218: 91, 140, 204

Page 219: $55 = 1^2 + 2^2 + 3^2 + 4^2 + 5^2$

$88 = 1^2 + 2^2 + 3^2 + 5^2 + 7^2$

CHAPTER 56

Page 223: Well, the triangular numbers are 1, 3, 6, 10, 15, 21, 28, 36, 45, 55 …

So the tetrahedral numbers are 1, 4, 10, 20, 35, 56, 84, 120 …

If you'd like to see a completely wicked diagram of a tetrahedral number, that you can spin around and look at from all sorts of angles, go to the fantastic Wolfram page: http://mathworld.wolfram.com/TetrahedralNumber.html

CHAPTER 57

Page 227: $8 = 2^2 + 2^2$, $17 = 3^2 + 2^3$, $54 = 3^3 + 3^3$, $32 = 4^2 + 2^4$, and my favourite: $6^2 + 2^6 = 100$

CHAPTER 61

Page 242: $x = 73$ gives $73^2 - 37y^2 = 1$, so $37y^2 = 73^2 - 1 = 5328$ and $y = 12$.

Page 243:

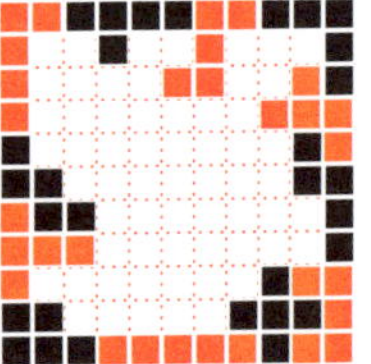

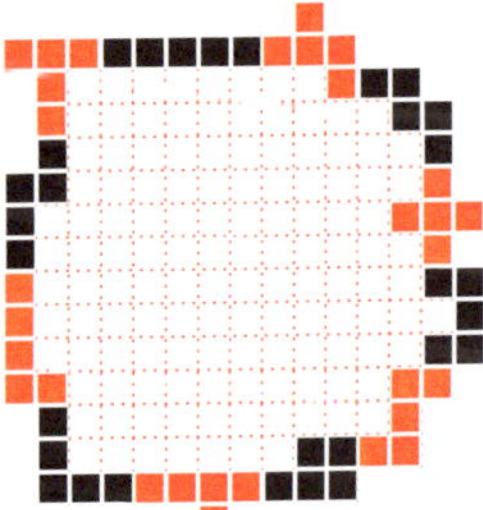

CHAPTER 62

Page 246: You can't arrange the 31 dominoes that way. When you remove the opposite corners you remove two squares of the same colour. This leaves 30 squares of one colour and 32 of the other. But 31 dominoes always cover 31 dark and 31 light squares.

CHAPTER 65

Page 258:

1	25	19	13	7
14	8	2	21	20
22	16	15	9	3
10	4	23	17	11
18	12	6	5	24

17	24	1	8	15
23	5	7	14	16
4	6	13	20	22
10	12	19	21	3
11	18	25	2	9

23	6	19	2	15
10	18	1	14	22
17	5	13	21	9
4	12	25	8	16
11	24	7	20	3

25	13	1	19	7
16	9	22	15	3
12	5	18	6	24
8	21	14	2	20
4	17	10	23	11

Page 260: $65 = 1^2 + 8^2 = 4^2 + 7^2$

$65^2 = 16^2 + 63^2 = 25^2 + 60^2 = 33^2 + 56^2 = 39^2 + 52^2$

CHAPTER 66

Page 262: 171 and 595. Also: 1 + 2 + 3 + 6 + 11 + 22 + 33 = 78. And, since 78 is greater than 66, 66 is therefore abundant.

CHAPTER 67

Page 267: $67^2 = 4489$ and 44 + 89 = 133

$67^3 = 300{,}763$ and 30 + 07 + 63 = 100

$67^4 = 20{,}151{,}121$ and 20 + 15 + 11 + 21 = 67

67 is a Kaprekar number of the fourth power.

CHAPTER 68

Page 272: 1, 7, 10, 13, 19, 23, 28, 31, 32, 44, 49, 68, 70, 79, 82, 86, 91, 94, 97, 100

CHAPTER 69

Page 274: 63 = LXIII and 12 + 24 + 9 + 9 + 9 = 63

Page 275: $69^2 = 4761$ and $69^3 = 328{,}509$. The answers use all the digits 0 to 9 once and once only.

CHAPTER 70

Page 279: 4900 or 70^2.

CHAPTER 71

Page 283: $m = 71$, $n = 7$.

CHAPTER 72

Page 286: 72 = 13 + 17 + 19 + 23 = 5 + 7 + 11 + 13 + 17 + 19

CHAPTER 73

Page 291: $78{,}557 \times 2^3 + 1 = 628{,}457 = 73 \times 8609$

CHAPTER 75

Page 298: 1, 6, 18, 40, 75, 126, 196, 288, 405, 550

Page 299: 1, 3, 5, 7, 9, 20, 31, 42, 53, 64, 75, 86, 97

CHAPTER 77

Page 306: $4^2 + 5^2 + 6^2$ or $2^2 + 3^2 + 8^2$

Page 307: 27, 35, 51, 57, 65, 77, 87, 93, 95

CHAPTER 79

Page 314: The cousin primes from 1 to 100 are (p, p + 4) for p = 3, 7, 13, 19, 37, 43, 67, 79 and 97

CHAPTER 80

Page 318:

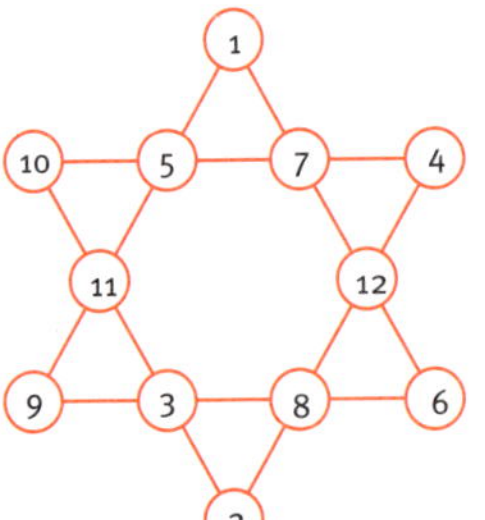

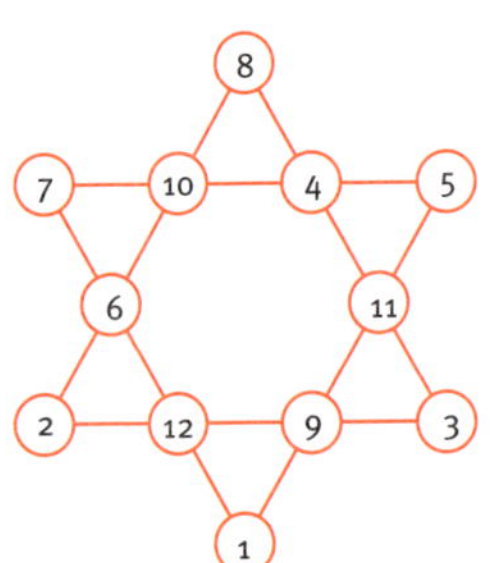

CHAPTER 83

Page 331: 5, 7, 11, 23, 47, 59, 83. Apart from 5 and 7 they are all equal to $12k - 1$ for some number k.

CHAPTER 84

Page 334: Let Diophantus live to D years of age. The riddle tells us that:

$$D = \frac{D}{6} + \frac{D}{12} + \frac{D}{7} + 5 + \frac{D}{2} + 4$$

$$D = \frac{(14D + 7D + 12D + 42D)}{84} + 9$$

$$D = \frac{75D}{84} + 9$$

$$84D = 75D + 756$$

$$9D = 756$$

$$D = 84$$

Page 336: 10, 12, 18, 20, 21, 24, 27, 30, 36, 40, 42, 45, 48, 50, 54, 60, 63, 70, 72, 80, 81, 84, 90, 100

CHAPTER 85

Page 340: $85 = 9^2 + 2^2 = 7^2 + 6^2$

CHAPTER 86

Page 342: 33, 34 and 35.

Page 344 (1):

$7 = 3 + 2 + 2 = 2 + 3 + 2 = 2 + 2 + 3$

$9 = 2 + 2 + 2 + 3 = 2 + 2 + 3 + 2 = 2 + 3 + 2 + 2 = 3 + 2 + 2 + 2 = 3 + 3 + 3$

$10 = 2 + 2 + 2 + 2 + 2 = 2 + 2 + 3 + 3$ and this combination of three 2s and two 3s can be ordered 6 ways.

Page 344 (2): It has no zeros in it. We've tested up to $2^{46,000,000}$ and not found another example, so 2^{86} is sometimes said to 'most likely' or 'almost certainly' be the highest power of 2 to contain no zeros.

CHAPTER 89

Page 355: 2, 3, 5, 11, 23, 29, 41, 53 and 83.

CHAPTER 90

Page 359: Using the recurrence relation, the Perrin numbers are 3, 0, 2, 3, 2, 5, 5, 7, 10, 12, 17, 22, 29, 39, 51, 68, 90 ...

CHAPTER 93

Page 371: 4, 6, 9, 10, 14, 15, 21, 22, 25, 26, 33, 34, 35, 38, 39, 46, 49, 51, 55, 57, 58, 62, 65, 69, 74, 77, 82, 85, 86, 87, 91, 93, 94 and 95

CHAPTER 94

Page 374: 4, 22, 27, 58, 85

CHAPTER 96

Page 383: 12, 18, 20, 24, 30, 36, 40, 42, 48, 54, 56, 60, 66, 70, 72, 78, 80, 84, 88, 90, 96, 100

CHAPTER 97

Page 388: To save space I'll write 97 (3,5) for $97 = 3 \times 2^5 + 1$, the values (a,b) you needed for the other Proth primes are 3 (1,1), 5 (1,2), 13 (3,2), 17 (1,4), 41 (5,3)

CHAPTER 98

Page 390 (1): You start with 99 kilos of water and 1 kilo of 'other non-water potato stuff'. The water is 99% of the potato mass and the 'stuff' is 1%.

But for the water to be 98% of the total mass you'd need 1 kilo of potato stuff and 49 kilos of water.

So the 100 kilos of potatoes have shrunk to weighing just 50 kilos!

Page 390 (2): $98 = 19 + 79$.

Page 392: $98 = 9^2 + 4^2 + 1^2 = 8^2 + 5^2 + 3^2$ and $98 = 9^2 + 3^2 + 2^2 + 2^2 = 8^2 + 4^2 + 3^2 + 3^2 = 7^2 + 6^2 + 3^2 + 2^2 = 6^2 + 6^2 + 5^2 + 1^2$

CHAPTER 99

Page 396:

Say, $x = 0.01010101010101 \ldots$

So, $100x = 1.010101010101 \ldots$

and $100x - x = 1.01010101 \ldots - 0.01010101 \ldots = 1$

So, $99x = 1$ and $x = \frac{1}{99}$

CHAPTER 100

Page 399:

$((1 \times 2) + 3) \times 4 \times 5 + 6 - 7 - 8 + 9 = 100$

$100 = 1 - 2 \times 3 + 4 \times 5 + 6 + 7 + 8 \times 9$

$100 = 1 - 2 \times 3 - 4 - 5 + 6 \times 7 + 8 \times 9$

$100 = 1 + 2 - 3 \times 4 - 5 + 6 \times 7 + 8 \times 9$

Index

ACKNOWLEDGEMENTS

This book is called *Adam Spencer's Big Book of Numbers* and it is my slightly cheesy photo on the front cover. As such, I take full responsibility for everything between the covers – especially any inaccuracies.

But I certainly didn't do all the hard slog myself. This book would not have been possible without the help of many other people to whom I am genuinely thankful.

A lot of the best non-mathematical facts in the book were uncovered by Belinda Middleweek and Laura Van den Honert in a frantic two-day web blitz. At various times I also relied upon the expertise of Professors Terry Gagen and Bob Armstrong, and one of the world's great scientific communicators, the redoubtable Dr Karl Kruszelnicki.

I spent a couple of fun and inspiring days with some very impressive young people who proofread the book – by that I mean tore it to pieces and helped me put it back together in a much better state. Emily and Sarah, Lola, Charlotte, Declan, Oliver, Alexander, Katherine, Spencer, Callum, Sam, Georgia and Angus, you are one bright packet of biscuits.

My wife Melanie trawled through my twitter account to find band names provided by some of my followers. My elder daughter Ellie helped for a brief period of time on the proviso that I put her name in the thanks ... and I'd also not want to upset my daughter Olivia so here she is as well. But seriously, guys, Dad wasn't around a fair bit for a month or two ... sorry.

To allow me to pursue my dream of independently publishing the *Big Book of Numbers*, the good people at Penguin, especially Andrea McNamara, released me from an ongoing arrangement and for that I respect their professionalism even more than when I loved working with them over a decade ago.

On the topic of awesome people at book-type places, this simply wouldn't have happened without the incredible work of three gurus at Xoum Publishing. Thank you so much Caroline Pegram for introducing me to Rod Morrison, Roy Chen and, in particular, Jon MacDonald, who made this book look better than I could ever have imagined. To RodMo, RayRoy and JMac, my eternal thanks, and apologies for the lame nicknames.